ISRAEL'S 2014 GAZA WAR
■■ A VIEW FROM TEL AVIV ■■

MARC SCHULMAN

©2015

ISRAEL'S 2014 GAZA WAR
▌▌ A VIEW FROM TEL AVIV ▌▌

Written by
Marc Schulman

Layout by Amy Erani

Published by **MULTIEDUCATOR INC**

180 E. Prospect Avenue • Mamaroneck, NY 10543

Email: marc@multied.com

ISBN # 978-1-885881-35-9

In loving memory of my parents,
who were responsible for instilling in me a great love of Israel,
though they never had the privilege of living here. – M.S.

❦

Table of Contents

Preface

A few weeks before the start of our most recent war with Gaza, I was poised to begin a new project. As an historian, the one story I have always longed to write is the history of the State of Israel. To call that endeavor a Herculean task would be a massive understatement and since I have a day job, I have deferred that undertaking for now. Instead, I am pursuing a slightly more manageable project: chronicling the History of Israel in the 21st century.

Since the Second Lebanon War, I had been writing a daily blog update on events in Israel, so I felt I had made a good start. By the time *Miv'tza Tzuk Eitan* (Operation Protective Edge) commenced, I had already written some twenty pages and calculated that within the year, the first draft would be complete. Then, the war began. This was not my first war in Israel, and not even my first experience with repeated alert sirens and rockets raining down (as I had been in Israel during the first Gulf War.) It was, however, the first war during which sirens sounded each and every day. This was the first war in which the children of my friends – and my children's friends – were fighting. (My daughter had completed her army service just six months prior).

It is much too early to write a true history of this war. There is much we do not know, and may not know for many years. I am neither an Arabist, nor an expert on Hamas. Therefore, this book will not attempt to present the Arab viewpoint. Rather, this is an attempt to look more closely at Israel over the course of the tumultuous events of July and August 2014. Wars generally bring out both the best and the worst in any society, and the same holds true for this most recent war. During the war I authored a daily column that summarized the military, political events of the war; and combined these with a more personal account of what it was like to live in Tel Aviv with a family during this time. That daily column was picked up by *Newsweek* who carried every one of my "Tel Aviv Diary" postings, from the fourth day of the war until its conclusion. Those reports comprise the core of this narrative and are presented here in full. Together with the additional background information and supplementary summaries reflecting on the effects on Israelis and Israeli society in the aftermath of this war, these accounts provide a basis for understanding Israel, circa – summer 2014.

Background

This is not an attempt to present a complete history of the Arab-Israeli conflict – that would require more than a full book. The intent of this narrative is to provide a concise history of Israel and Gaza. To do so, we must first start with some retrospective background on the relationship between Israel and Hamas.

The Gaza strip is 25 miles long, with an average width of 5 miles. There are 1.8 million people living in Gaza, most of whom are descendants of refugees from Israel's 1948 War of Independence. These Gaza residents, and the other Palestinian refugees, are the only group in the world whose descendants are recognized as refugees.

Israel's troubled history with Gaza extends back to ancient times, when Samson fought the Philistines there. In modern history, the Gaza Strip was the destination of many Palestinian refugees, escaping the battles of the War of Independence. During the War of Independence, Egyptian troops advanced all the way into Israel – reaching the area of what is today the city of Ashdod. There, the Egyptians were stopped. Eventually reaching Eilat, Israeli troops pushed the Egyptian army out of nearly the entire area of what had been mandatory Palestine. The one exception was the Gaza Strip, where the Egyptian army remained.

The Israeli War of Independence ended in 1949, with an armistice agreement signed on the island of Rhodes. At the war's end, the Gaza Strip held an indigenous population of 50,000, and an additional 120,000 refugees. The refugees were denied citizenship and had no employment prospects in Gaza. They lived solely off of support from the United Nations. The Palestinians in Gaza experienced the worst conditions of all the 1948 Arab refugees, with the bleakest future. Initially, the Egyptian government took control over the border. Within a short time, however, the Egyptians allowed – and in some cases, even encouraged – cross-border raids by Palestinians into Southern Israel. Those raids grew in frequency and severity. Twenty-six Israelis were killed or wounded by terrorist attacks in Southern Israel in 1952. In 1953, that number rose to 50. By 1955, it had reached 192.[1]

Traveling in Southern Israel was dangerous. Every *kibbutz*, along with the other settlements in the South, managed its own 'border'. Israel responded to these attacks with ever-stronger reprisal raids against Egyptian positions and personnel in Gaza. The U.N. condemned Israel's retaliatory raids – without condemning the terrorist attacks – in a pattern that Israelis today find very familiar.

In 1956, Israel joined with France and England in a military campaign against Egypt. For Israel, this was a preemptive attack, before Egypt (led at the time by Gamal Abdul Nasser) could receive a massive arms shipment from the Soviet Union. This was also a chance for Israel to find solutions to the ongoing attacks launched from Gaza, as well as to the Egyptian blockade of the Straits of Tiran. In what was to become known as The Sinai Campaign, Israel rapidly conquered Gaza and the most of the Sinai desert. Though Israel was eventually compelled to withdraw from Sinai after strong U.N. resolutions supported by the United States, Israel did not initially withdraw from the Gaza Strip. It was holding out for a plan to end the attacks that had been initiated over the years. A solution was reached that called for the stationing of United Nations troops on the Gaza border, replacing the Egyptian troops previously stationed there.

The period between 1957 and 1967 was a quiet one on the Gaza border, until a series of events, (initiated by the Egyptian leader Gamal Abdul Nasser) came to pass. The upshot was Egypt's demand that the U.N. remove its troops from Gaza and from the Straits of Tiran. For reasons that have never been fully explained, U.N. Secretary General U. Thant complied immediately. As a result, Egyptian troops entered Gaza and Sinai. They closed the Straits of Tiran to Israel and threatened to attack. Israel responded with a lightning preemptive strike that resulted in the capture of Sinai and the Gaza Strip from Egypt. In a matter of just six days the map of the Middle East had changed. In addition to the entire Sinai Peninsula and the Gaza Strip, Israel stood in control of the West Bank (formerly held by Jordan) and a chunk of Syria, called the Golan Heights. Israel proceeded to impose a military administration on Gaza. For the first few years of the military governance of the Strip, Israel faced significant resistance. That resistance subsided somewhat by early 1970, after General Ariel Sharon, (head of the Israel Defense Force's Southern Command) turned his attention to pacifying the strip.

In 1970, Israel established *K'far Darom*, its first *Nahal* settlement in Gaza. *Nahal* settlements are communal installations founded by the military, made up of youngsters who choose to remain on the settlement after their army service. The goal of the settlements in Gaza was to divide the Strip in half, using the settlements as a barrier between the two areas. Israel's Gaza settlements grew slowly, only gaining a significant number of new members when Israel pulled out of Sinai in 1982, a move that served to force the Sinai settlers to do the same.

▲ Yitzchak Rabin, Bill Clinton, and Yassir Arafat at the
Oslo Accords signing ceremony on 13 September 1993

But resistance continued against Israeli control though that resistance was held in check by the Israeli military. By 1987, tension was rising in the Gaza Strip. On December 8th, an Israeli truck driver lost control over his truck near the *Jabaliya* refugee camp. The driver plowed into a car and killed four locals. Rumors spread that this was a deliberate attempt to kill Arabs. Rioting broke out almost immediately in Gaza. A besieged local army attachment responded with live fire, killing one demonstrator and wounding thirty. Thus began what became known as the First *Intifada* or uprising. The *Intifada* immediately spread to the West Bank. Between December 1987 and April 1990, 1,054 Palestinians lost their lives. During the same period 100 Israeli civilians and 60 soldiers were killed.

The first *Intifada* continued until the signing of the Oslo Peace agreements between the Palestinian Authority and Israel in September 1993. According to the agreement, Israel was to pull out of all of Gaza – with the exception of the Jewish settlements. Israel was to maintain control of the crossings into Gaza, including the approaches from the sea. The Palestinian areas were to be demilitarized, allowed only to maintain a small police force, without advanced weaponry.

But in 1987, a new force was heard from for the first time – *Hamas*. Translated from the Arabic, *"Hamas"* implies a devotion to the path of *Allah*. It is, however, also an acronym for *"Harakat al-Muqāwama al-Islāmiyya'*, which means, "Islamic Resistance Movement." *Hamas* was a direct descendent of Egypt's Muslim Brotherhood, which had been founded in 1928. In 1988, *Hamas* issued its charter, which states: *"Allah* is its goal, the Prophet is the model, the *Qur'an* its Constitution, *Jihad* its path, and death for the sake of *Allah* its most sublime belief." *Hamas's* charter goes on to add: "Zionist invaders, the Day of Judgment will not come until Muslims fight the Jews [and kill them]; until the Jews hide behind rocks and trees, which will cry: 'Oh Muslim! There is a Jew hiding behind me, come on and kill him!'"

In 1989, *Hamas* took its first aggressive actions, kidnapping and murdering two Israelis soldiers. In 1991, *Hamas* established its military wing, the *Iz ad Din al Qassam* Brigades. The *Hamas* military brigades carried out repeated attacks against Israelis, including civilians. *Hamas* claimed their actions were in retaliation for the killing of Palestinian civilians.

Hamas opposed the negotiations between Israel and the PLO that led to the Olso agreement. They believe that all of the land of Palestine is sacred land, thereby no compromise is possible. As negotiations proceeded, *Hamas* stepped up their attacks inside Israel. Between 1993 and 1994, *Hamas* carried out three-dozen terrorist attacks on Israeli soil – including a suicide bombing on a bus in Tel Aviv that killed 22 passengers, and an attack at a bus stop near Netanya in which 21 soldiers were killed. Negotiations proceeded despite these attacks. On

September 28th 1995, Oslo II was signed. Among its requirements was the withdrawal of Israeli forces from all major Palestinians cities.

On November 4th 1995, Prime Minister Yitzhak Rabin was assassinated by Israeli extremist, Yigal Amir, who opposed any compromise with the Palestinians. Shimon Peres, considered by many the architect of the Oslo accords, became the new Prime Minister. *Hamas* remained determined to derail the peace process. In 1996, *Hamas* stepped up its bombing campaign carrying out two bus bombings in Jerusalem that killed 45, and a bombing at *Dizengoff* Center in Tel Aviv that caused the deaths of an additional 13 Israelis. After the first Jerusalem bombing, the Israeli secret service killed *Yahiya Ayash*, the chief bomb maker for *Hamas*.

In 1996, Benjamin Netanyahu won the elections against Shimon Peres – in large measure, because of the *Hamas* bombings. While Netanyahu went on to sign the Wye Accords, giving the Palestinians greater control over parts of the city of Hebron, Netanyahu had been a vocal opponent of the Oslo accords and did all he could to stop that process. In the 1999 elections, Netanyahu was defeated by Ehud Barak, who became Prime Minister with a mandate to reach a final peace agreement with the Palestinians and the Syrians. After failing to come to terms with the Syrians, Barak tried to achieve an accord with the Palestinians at Camp David. Those efforts failed when Palestinian leader Yasser Arafat turned down Barak's offer and did not respond with a counter-offer.

Soon after, the second *Intifada* was launched. *Hamas* was an active participant in the many bombing attacks that were carried out, though it was but one of many players. These attacks were implemented primarily by bombers based in the West Bank, who had easy access to the rest of Israel. There were a number of attacks in Gaza, but these were more difficult targets than the undefended targets inside Israel. The second

Intifada resulted in severe restrictions being placed on Palestinian movement from the West bank to Gaza, as well as from Gaza into Israel proper.

Ariel Sharon was elected Prime Minister in 2001. In September, the first *Qassam* rockets were fired at *Sderot* from Gaza. In March 2002, after a particularly deadly bombing in Netanya on the first night of Passover, Sharon ordered "Operation Defensive Shield", which included reoccupation of the major Palestinian cities to root out terrorism. The strategy was only partially successful. However, combined with the building of the separation fence, terrorism was significantly reduced. The death of Yasser Arafat, who had supported the "armed struggle", coupled with the shift to leadership status of his successor, Mahmoud Abbas, brought an end to the second *Intifada*.

In February 2004, Ariel Sharon surprised Israel – and the world – by announcing a unilateral withdrawal from the Gaza Strip. Sharon became convinced that a political solution was not attainable, while realizing that the political and military costs of sustaining the occupation were too high. During this period, Israel was losing soldiers to attacks in Gaza on a regular basis – especially in the area known as "The Philadelphia Corridor" (the area along the border with Egypt), in which Gazans were building tunnels and smuggling weapons, together with other supplies into the Strip. Sharon announced that Israeli presence in Gaza would end by the summer of 2005, and that any Israeli settlers who did not leave voluntarily would be forcibly removed.

Still, the attacks by *Hamas* did not end. Over the course of March 2004, ten Israelis were killed during a bombing in Beersheva. Israel retaliated by attacking *Hamas* positions in refugee camps inside Gaza. Later that same month, Israel assassinated the spiritual and political leader of *Hamas*, Sheik Ahmed Yassin.

In January 2005, Mahmoud Abbas won the election to serve as President of the Palestinian Authority. Abbas was elected on a platform of ending the violence and seeking a political solution with Israel. *Hamas* did not go along with Abbas' plan and continued to fire rockets into Israel. The Palestinian police had only limited success in trying to stop the attacks.

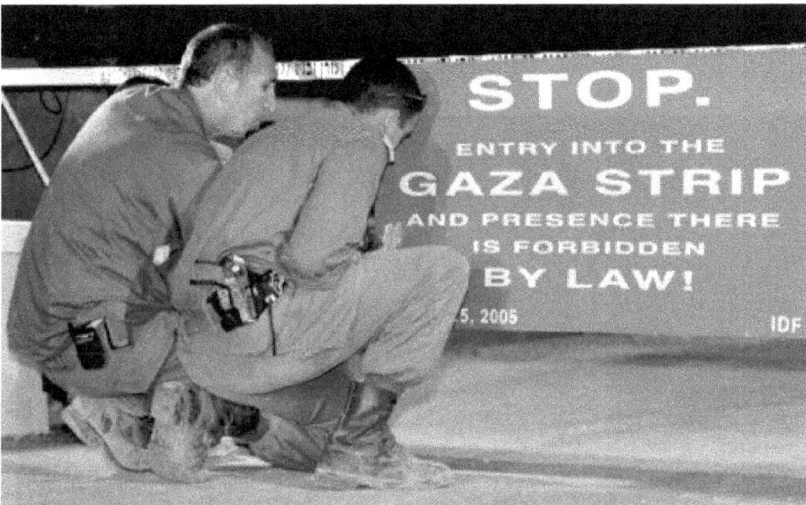

▲ **August 14, 2005. IDF forces and the Israeli Police close the Kisufim checkpoint to citizens by the order of the Prime Minister, Ariel Sharon, and the Ministry of Defense.**

In August 2005, Israel forcibly removed the 9,000 Jewish settlers living in Gaza. As of September 12th 2005, Israel had completed its withdrawal from Gaza, turning the administration of the Strip completely over to the Palestinian Authority. Israel continued to supply water and electricity to Gaza, and of course, maintained control over all of the border crossings into Israel. The border crossing into Egypt at Rafah was supervised by the European Union, and the "Border Assistance Mission for the Rafah Crossing (EU BAM Rafah)", who were supposed to ensure no weapons were allowed into Gaza.

In January 2006, *Hamas* was the key victor in the Palestinian Parliamentary elections. *Hamas* won 74 seats in a 132 parliament by gaining 44.45 % of the vote. The United States and other members of what was called the Quartet (US Russia, France and UK) demanded that *Hamas* accept the existence of Israel and all previous agreements that had been reached with Israel. *Hamas* refused to comply. As a result, both the Quartet and Israel officially refused to deal with *Hamas*.

▲ **Smuggling Tunnel, Rafah, Gaza Strip**

Via a tunnel, *Hamas* attacked an Israeli position outside of the Gaza Strip on June 25th 2006. During the course of that attack, Israeli soldier, Gilad Shalit was kidnapped. In response to the kidnapping, Israel launched "Operation Pillar of Cloud," which failed to obtain Shalit's release.

A full-scale civil war broke out in Gaza between the forces of the Palestinian Authority and *Hamas* in June 2007. This resulted in an absolute victory for *Hamas*. *Hamas* gained control of all the institutions controlled by the Authority, and took complete charge of Gaza.

Hamas and Israel continued to battle it out in the form of missile fire by the Qassam Brigade at Israel. A six-month ceasefire was negotiated between *Hamas* and Israel. Much to Israel's surprise, when the ceasefire ended in December 2008, *Hamas* chose not renew it. Instead, *Hamas* unleashed massive rocket attacks on Southern Israel. Israel responded with "Operation Cast Lead", beginning with an opening salvo air assault on December 27th, in which many members of *Hamas* were killed. Israel began a major ground assault on Gaza on January 3rd, to try to end the rocket fire and attempt once more to free Gilad Shalit. Just over two weeks later, on January 18th, Israel announced a one-sided ceasefire and withdrew. Over the course of this campaign, approximately 1,300 Palestinians were killed, including 700-800 *Hamas* fighters. Israel lost 11 men. Israel was criticized by the United Nations and others, who claimed Israel was targeting civilians. Israel strongly denied the charge and refused to cooperate with the U.N. investigatory "Goldstone Commission", which was highly critical of Israel.

In November 2012, Israel responded to an increase in missile fire from Gaza by launching "Operation Pillar of Defense". The Operation began with the aerial assassination of Ahmed Jabari, the head of Hamas's military wing. Israel attacked 1,500 targets in the Gaza, in an eight-day campaign. Over 130 Palestinians were killed, including 53 civilians. *Hamas* and *Islamic Jihad* fired a total of 1,456 rockets at Israel during the operation, including the first missiles launched at Tel Aviv. This was also the first time the Israeli *Iron Dome* anti-missile system was put into service. Iron Dome successfully intercepted most of the missiles. Egyptian President Morsi negotiated a ceasefire that called for a halt all hostilities, by both sides, and would lead to the opening of the border crossing.

In June 2013, a counter-revolution took place in Egypt and Mohammed Morsi, the Muslim Brotherhood President, was ousted.

He was replaced with a member of the Egyptian military. The new government in Egypt began a crackdown on the smuggling of goods via the tunnels under the Gaza Egyptian border. The Egyptian army later closed the border crossing at *Rafah,* following violence against in its soldiers in Sinai.

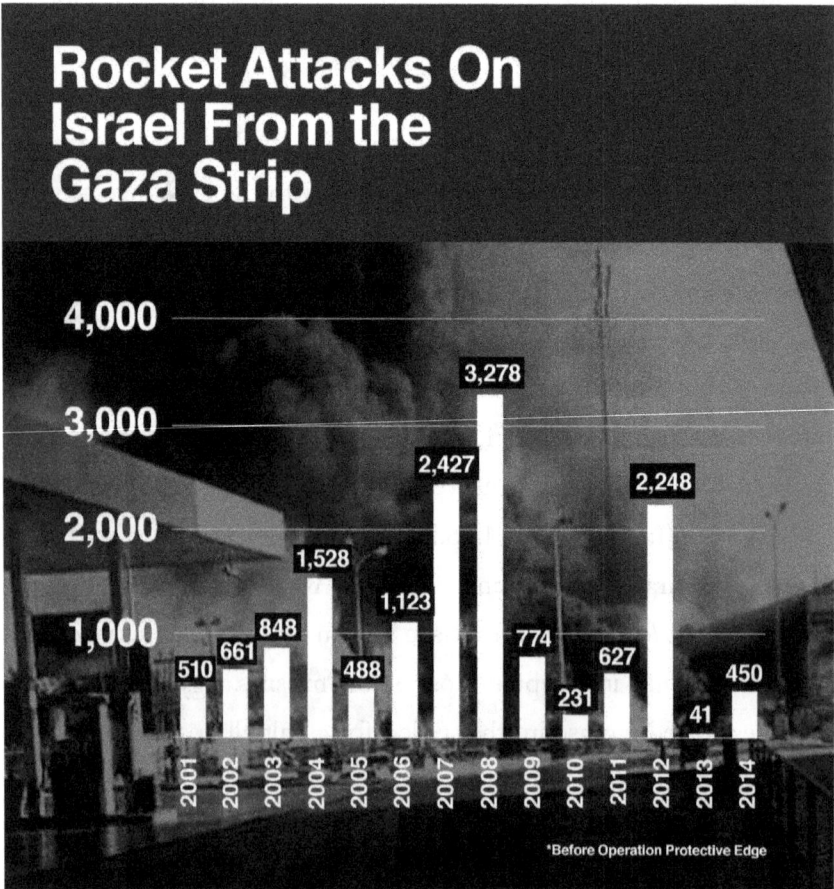

Rocket Attacks On Israel From the Gaza Strip

4,000

3,000

3,278

2,427

2,248

2,000

1,528

1,123

1,000

848

774

661

627

450

510

488

231

41

2001
2002
2003
2004
2005
2006
2007
2008
2009
2010
2011
2012
2013
2014

*Before Operation Protective Edge

The Year 2014

It is ironic, considering what was to follow, but the first major event in Israel in 2014, was the death of former Prime Minister Ariel Sharon. 'Arik' Sharon, architect of the unilateral withdrawal from Gaza, had been in a coma for six years. On January 4th, he died and a State funeral followed. His death triggered a time of reflection. Six years after the withdrawal from Gaza, it was clear opinions in Israel had not changed: those who opposed the unilateral withdrawal could point to the fact that Hamas had come to power and Israel had already fought two small wars with Hamas in Gaza. Those who had supported the withdrawal could point out with equal fervor that Israeli soldiers were no longer dying in Gaza from either defending the settlements or trying to patrol the Philadelphia corridor. Furthermore, thousands of soldiers were no longer serving inside Gaza. But no one reflecting on Ariel Sharon's life and legacy early in January 2014 could have predicted that that summer, over half the population of Israel would be running to shelters as a result of missile barrages being launched from Gaza.

By early 2014, the Iranian issue – one of the central concerns casting a long shadow over Israel for the better part of the previous two years – had receded a little. In January, Iran and the Western Powers (P+5) reached an interim agreement, under which Iran would temporarily restrict its nuclear development, for which, in return the West would ease some of its sanctions. For most of the rest of the year, negotiations would take place between Iran and the Western powers to try to reach a permanent agreement. While no one in Israel believed this had solved the difficulties with Iran, and many believed this was just another effort on the part of the Iranians to gain time, what was clear was that (at least for the moment), an Iranian–Israeli confrontation had been avoided for the time being.

As winter turned into spring, two sets of unrelated events dominated Israel's media. First, a series of ongoing scandals and trials of former and current Israeli politicians headlined the news followed by the attempt of U.S. Secretary of State John Kerry to broker an agreement between Israel and the Palestinians. Kerry's talks had been going on for nearly a year, and while there had never been high expectations for the success of the negotiations, it was becoming clear by March that whatever small chance of success had existed was virtually gone.

In early March, Prime Minister Netanyahu traveled to the United States to address the annual AIPAC conference. President Obama had declined an invitation to address the gathering. Before the scheduled pre-conference bilateral meeting between Prime Minister Netanyahu and President Obama, Obama was interviewed by Jeffery Goldberg of Bloomberg News. In that interview, President Obama stated:

> If you see no peace deal and continued aggressive settlement construction – and we have seen more aggressive settlement construction over the last couple years than we've seen in a very long time – if Palestinians come to believe that the possibility of a contiguous sovereign Palestinian State is no longer within reach, then our ability to manage the international fallout is going to be limited.[2]

The interview caused considerable consternation in Israel, especially by those who believed Netanyahu was being unfairly singled out for criticism. A day after the interview was posted online, (but four days after it had taken place), Netanyahu and Obama met at the White House. Reports indicate it almost seemed as if the Goldberg interview (and the message contained within it) had never taken place. The meeting was said to be rather cordial. But between the time Obama gave the interview and his meeting with Netanyahu, the world had changed in an unexpected way: Russia had invaded Crimea.

During their meeting, President Obama reportedly promised Netanyahu he would pressure Palestinian Authority President Abbas to accept the Framework Agreement that Secretary of State Kerry was preparing and agree to extend the talks. The following day, Netanyahu spoke at the AIPAC conference. While Netanyahu took his usual hard line approach to Iran, when it came to the West Bank and peace with the Palestinians, he was uncharacteristically optimistic. As Barak Ravid wrote in *Ha'Aretz*:

> If anyone needs to be concerned about the address Prime Minister Benjamin Netanyahu delivered at AIPAC's annual policy conference on Tuesday, it's the Council of Jewish Communities of Judea and Samaria and the Settlers' Caucus in the *Knesset*.[3]

Netanyahu stated in his speech:

> I'm prepared to make a historic peace with our Palestinian neighbors; a peace that would end a century of conflict and bloodshed. Peace would be good for us. Peace would be good for the Palestinians. Though peace would also open up the possibility of establishing formal ties between Israel and leading countries in the Arab world.

If one just read the words and did not know who was speaking, one could assume it was Shimon Peres, and not Benjamin Netanyahu.

Whatever limited optimism existed after Netanyahu's visit to Washington, seemed to evaporate the next day when the Palestinian Authority responded by stating they had no more concessions to make.

In the midst of all this, a mini-crisis developed in U.S.-Israeli relations, when Defense Minister Ya'alon publicly called the United States 'weak' in its response to Crimea, and downplayed its reliability as an ally. U.S. officials were furious at Ya'alon and demanded an apology; which Ya'alon vaguely delivered, but not really to the satisfaction of the American administration.

On March 17th, Mahmoud Abbas met with President Obama at the White House. President Obama's definitive goal for the meeting was to get Abbas to agree to the Kerry's framework. But that is not what happened. By all accounts, Abbas refused to accept three key aspects of the framework: Netanyahu's demand that Israel be recognized as a Jewish State; accepting that refugees would not return to pre-1967 Israel; and any agreement would represent the end of conflict. While Netanyahu's request that the Palestinians recognize Israel as a Jewish state was controversial, the other parts are considered prerequisites by almost all Israelis across the political spectrum, in order for there to be any further substantial Israeli concessions. The United States was unwilling to publish the framework, nor were they willing to clearly state that it was Abbas who refused to accept key aspects of the proposal.

By the next day, the Israeli government was coming to the realization that no framework deal was going to be accepted and that the Palestinians might not extend the talks. Justice Minister Tzipi Livni, who was also Israel's key negotiator, warned *"the prison keys remain the hands of Mahmoud Abbas, and the decisions he makes in the next few days"* [4] Livni was referring to the final cohort, of a four-part prisoners release program that Israel had agreed to implement, as an incentive for the Palestinians to enter into negotiations. The release of this particular group of prisoners was extremely unpopular in Israel – since it included prisoners who had killed Israelis. Israel had insisted that the releases take place in phases, the fear being that Israel would release the prisoners, and then the Palestinians would walk away from the talks. Now, towards the end of a full year of the negotiations, the last group of prisoners was due to be released. Israel was worried that if the release was affected, the Palestinians would announce a few days later they were not going to resume the talks. In the eyes of the Palestinians, however, the release of the final group of prisoners was due them. They considered it part of the deal that had allowed the talks to go on for a year.

It was clear to all observers that the talks were on the verge of a collapse. The Americans, led by Secretary of State Kerry, began a marathon session to try to find a way to extend the talks. Various ideas were floated. One of the suggestions included the release of Jonathan Pollard, the American Jew convicted of spying for Israel. Pollard has been in jail for over a quarter of century, and Israelis believe his release is long past due. A deal was in the offing that would initiate a freeze on settlement construction along with the release of additional prisoners, in return for the Palestinians continuing the negotiations, and the U.S. releasing Pollard. It looked as if a deal was imminent. The Israeli cabinet was poised to vote on the agreement, in what promised to be a stormy meeting.

However, to everyone's surprise, the Palestinians decided instead to further their bid for official statehood, by asking to be recognized by various U.N. agencies – this move being in direct conflict with their agreement not to seek U.N. recognition during the year that the negotiations would continue. This threw a monkey wrench into all of the efforts to restart the Israeli-Palestinian negotiations. To further complicate matters, Israeli Minister of Housing, and member of the right-wing *HaBayit HaYehudi* party announced a tender offer for housing in East Jerusalem. The tender was not new, but was, in fact, just an attempt by Uri Ariel to further cripple the talks. It remains unclear why Abbas decided to announce that the Palestinians were going to approach the United Nations for recognition at this time, thereby torpedoing any chance of the talks resuming. One possible explanation offered was that the Palestinians were upset that the U.S. was offering up Pollard as a gift to the Israelis to continue the talks and agree to the prisoner release, while the Palestinians were not getting anything new.

On April 2nd, the Israelis and Palestinians held a very acrimonious meeting, in which each side accused the other of sabotaging the talks. It looked like the negotiations had reached an end. By the next day, sources in Secretary of State Kerry's entourage were beginning to say that the US had reached the limits of what it could achieve, and if the parties were not willing to reach an agreement, there was nothing more the U.S. could do. Still, it seemed both sides were not yet willing to walk away from the talks. Efforts continued to find a formula that would allow the negotiations to continue. The Palestinians indicated that if Israel were to release the fourth group of prisoners, they would hold off submitting their requests to the U.N., and consider returning to the negotiating table. In the meantime, both sides played the blame game. The Palestinians claimed the Israelis had torpedoed the talks, by not releasing the fourth group of prisoners, and by announcing additional building in the East Jerusalem. At the same time, the Israelis blamed the Palestinians for not agreeing to the framework agreement and by taking the unilateral action of going to the U.N.

Attempts continued to navigate toward an extension, as the official deadline for the end of the talks neared. Martin Indyk, Secretary of State Kerry's representative met with both sides and tried to get the talks going. The breakdown of negotiations was taking place against the background of significant turmoil in the world. In the Middle East, the Syrian Civil War had entered its third year; and the growth of a new element, the extremist Islamic opposition group *"Daesh"* (Known in English as ISIS), which was rapidly expanding its control over parts of Syria, and Iraq as well. The actions of *Daesh* were suddenly grabbing the world's attention, a world that was utterly stumped as to how to respond. In the Ukraine, it seemed as if the Cold War had resumed. After Ukrainians overthrew a leader, though democratically elected, seemed to clearly be taking his marching orders from the Kremlin; in a naked display of power, the Russians seized Crimea (which was part of the Ukraine), claiming this action was in response to the will of the Crimean people. That crisis did not look like it would be ending any time soon. So, as the Americans were scrambling to find a way to save the Israeli-Palestinian talks, problems in other parts of the world appeared to be much more urgent.

The final nail in the coffin of the most recent round of Israeli-Palestinian peace talks occurred on April 25th, when (seemingly out of the blue), the Palestinian Authority announced they had reached an agreement with Hamas to form a unity government. While Hamas and the Palestinian Authority had been speaking about joining for some time, the actual announcement caught both the United States and Israel by surprise. For Israelis, this was effectively the end of the peace talks. There was no point in moving forward if the Palestinians embraced Hamas – a group that had never renounced their charter, and has refused to accept the conditions for recognition set by the U.S., the U.N., the E.U. and Russia (the quartet). For Israel, and to a lesser extent, the United States, the Palestinian Authority-Hamas agreement was a backdoor strategy to legitimize Hamas, without the need for the group to accept the conditions set forth either by the quartet or by Israel, the most important of which being acceptance of the previous agreements between Israel and the Palestinians. The Palestinian Authority claimed there was no problem, since the new joint government would recognize Israel. At a Press Conference with South Korean President Park on April 25th, President Obama expressed his frustration:

> And the fact that most recently President Abbas took the unhelpful step of rejoining talks with Hamas is just one of a series of choices that both the Israelis and the Palestinians have made that are not conducive to trying to resolve this crisis. And I make no apologies for supporting Secretary of State Kerry's efforts – tireless efforts – despite long odds, to keep on trying to bring the parties together.
>
> There may come a point at which there just needs to be a pause and both sides need to look at the alternatives. As I've said in the past and I will continue to repeat: Nobody has offered me a serious scenario in which peace is not made between Israelis and Palestinians and we have a secure, democratic Jewish state of Israel and the Palestinians have a state. Folks can posture; folks can cling to maximalist positions; but realistically, there's one door, and that is the two parties getting together and making some very difficult political compromises in order to secure the future of both Israelis and Palestinians for future generations.[5]

At this point, all efforts to revive the talks were dead. Martin Indyk, the indefatigable emissary, resigned his post as Middle East Special Envoy, stating there was not much more he could do at this time.

For a month, there was much hand-wringing over why the peace talks collapsed, and why the Palestinians decided in favor of a unity government at this time. Most analyses pointed to the perilous condition in which the Hamas found itself, i.e., unable to pay salaries and totally isolated politically. Hamas decided the only way out of their crisis was to join a unity government. As such, they were willing to make the concessions necessary to make that government viable. On June 2nd, an official agreement on unity was signed. On paper, the unity agreement looked real. Whether such a government could actually be implemented was very much open to doubt.

For a short time thereafter, Israel was distracted by the question of who would become its next President. President Peres was reaching the end of his seven-year term in June. The one candidate who most passionately wanted the job – former speaker of the Knesset Reuven "Ruby" Rivlin, who was a Likud member, and thus a natural candidate for the job – had had a stormy relationship with Prime Minister Netanyahu. Netanyahu was known to oppose Rivlin's candidacy, and was actively searching for an alternative. The Labor party's major alternate contender, former General and Cabinet Minister Binyamin "Fuad" Ben-Eliezer was suddenly engulfed in a police investigation, and forced to withdraw from the race. A number of a non-politicians, including Dan Shechtman, Nobel Prize winning professor from the Technion; as well as former Supreme Court Justice Dalia Dorner, competed for the Presidential post. In the end, the only viable alternative candidate was Meir Sheetrit, former Minister of Justice, and current member of Tzipi Livni's *HaT'nua* party. It seemed as if Sheetrit was on the cusp of victory when Rivlin managed to strike a deal with the Ultra-Orthodox bloc and as a result, was elected President.

On the night of June 12th, three Jewish teenagers waiting for rides home were kidnapped outside of Gush Etzion, on the West Bank. The three young men: Gilad Shaer, Naftali Fraenkel, and Eyal Yifrach entered a car at 10:25PM. Moments later they realized their mistake, and one of them managed to call the police and inform them they had been kidnapped. The kidnappers, Marwan Kawasmeh and Amar Abu-Isa, who had not planned to capture three people together, panicked and killed the students. They then destroyed the car they were traveling in and transferred to another car to dispose of the bodies.

Of course, none of this was known at the time and so a massive operation to locate and free the abductees began. That operation saw large number of Israeli troops surging into the Hebron area – where the kidnappers were thought to reside. On the political front, Hebron was also where the Israeli government directed the propaganda offensive against Hamas (who Israel claimed was behind the kidnappings). To a lesser extent, Israel also held the Palestinian Authority accountable for entering into a coalition government with Hamas.

For the next two weeks, Israel and World Jewry were gripped by the drama unfolding in Israel. Prayer vigils and Facebook campaigns were mounted and rallies held to *"Bring back our boys"*. The government arrested hundreds of Hamas members in the West Bank, including most of those who had been released as part of the Gilad Shalit prisoner exchange. All this was going on; despite the fact that the government knew the three teens were almost certainly dead. Forensics from the abandoned car yielded evidence that the three had been killed shortly after being abducted.

On the night of June 29th, thousands packed Rabin Square in a rally in solidarity with the families of the abducted teens. Normally, events in the West Bank seem far and removed to the average resident of Tel Aviv, but the kidnapping of the students resonated strongly. Parents and grandparents across the country had the horrifying feeling that this could have happened to their kids, if kidnappers had had the opportunity. Of course there are very few residents of Tel Aviv that would ever allow their teenagers to hitchhike in the West Bank, but still … the very next day the boys' bodies were found.

▲ Graves of the three teenagers (l-r Eyal Yifrach z"l, Naftali Fraenkel z"l, and Gilad Saher z"l) in the Modi'in Cemetary

Throughout Israel, people reacted with profound sorrow. The country's identification with the family of the slain youths was strengthened and amplified by the quiet dignity exhibited by the bereaved families in public and in private. Tens of thousands attended the funerals and the burials, which were held together.

The death of three students unleashed an unprecedented wave of hatred against Arabs in Israel – an example of which was the vitriolic call by the head of *B'nei Akiva*, Rabbi Noam Perl, on his Facebook page: *"An entire nation and thousands of years of history demand revenge,"* Perl went on to say:

> The government of Israel is gathering for a revenge meeting that isn't a grief meeting. The landlord has gone mad at the sight of his sons' bodies. A government that turns the army of searchers to an army of avengers; an army that will not stop at 300 Philistine foreskins.[6]

Rabbi Perl's voice was not a lone voice demanding revenge. The saga of the search had created an emotional bond with the abductees, far stronger than the bond forged with the average victim of terror attacks. Throughout the crisis, the Palestinian Authority fully cooperated with the Israeli army in trying to find the kidnapers. Furthermore, President Abbas publicly attacked the kidnappings in Arabic, in the Arab media.

▲ Mourning tent of Muhamed Abu Khdeir, Jerusalem 2014

The anger and blood lust for revenge led to its natural, though previously unimaginable conclusion, when on July 2nd, the body of 16-year-old Mohammed Abu Khdeir was found badly burned in the Jerusalem Forest. Khdeir had been kidnapped by three Israelis and was burned alive, in retribution for the killing of three Israeli students. Israelis were in shock. They found it hard to believe that some of their own could actually kill – and not only kill, but burn alive an innocent Palestinian teenager – in retribution for the killing. While not as traumatic as the assassination of Prime Minister Rabin, for a day, it seemed that this hatred-filled killing might actually kindle a serious national debate on hate in the society. Although some discussion ensued on the national level about this unprecedented act, the ensuing war with Hamas served to put everything else on hold.

It is impossible to determine the exact moment when the war began – was it when the first rocket was fired from Gaza into Israel? Or, was it when Israel began retaliating? If forced to choose one instigating moment, I would point to the evening of July 7th – the night Hamas fired 40 rockets, within the span of one hour, at cities as far north as Ashdod and Rehovot. That attack guaranteed Israel would no longer be able to respond symbolically, but would instead be forced to respond in kind. After that evening, the war was on.

June 22, 2014
Problems on Three Fronts -
ISIS and Israel

The ongoing turmoil in the Arab world impacted Israel today on three of its four problematic fronts. Today, there was an attempt by a terrorist to enter a Kibbutz in the south. How he got there from Gaza is not at all clear. Thankfully, he was caught. The search in the West Bank continues for the kidnapped students, while at the same time doing everything possible to undermine Hamas there. Tonight the sense is that this effort has gone as far as it can, and will be called off in the next day or two. Meanwhile, the search continues. The official word is that the boys are alive. However, with much of the search taking place in water wells and similar places, you can draw your own conclusions.

On our Northern border an anti-tank missile was fired at an Israeli car. The explosion killed the vehicle's 15-year old occupant who had gone to work with his father this morning – the first day of summer vacation. It's not totally clear which forces are responsible– those aligned with Assad (i.e. Hezbollah) or the Syrian Rebels. Either way this represents a dangerous escalation. Although the real concern at the moment is that the ISIS in Iraq has ignited something that could easily spread throughout the Arab world – and first and foremost to Jordan. Their success in Iraq seems contagious. Of course, before it can spread back to Jordan, Lebanon, and of course our way ISIS first has to contend with the Shiites who are mobilizing. It's hard to say that we are uninterested in the outcome. Though it is even harder still to say that we could possibly have rational policy. The general rule of thumb is that when your enemies are killing each other – stay out of the way. However, there are no assurances that when all this sorts itself out we will not find ourselves in a worse position.

▲ Factory bursts into flames after rocket attack on Sderot on 6/28/14

June 29, 2014
Rocket Fire, Rally, Supreme Court Ruling, and More

It's been an incredibly busy news day in Israel. It is unclear what is even the most important piece of news. So I guess I will start with the latest update. Two more sets of rockets were fired on Southern Israel tonight. Iron Dome intercepted those missiles that might have landed on populated areas. Yesterday we were not so lucky. A missile hit a factory in S'derot. A secondary explosion destroyed much of the factory. Fortunately, no one was seriously hurt. It seems the rockets are being fired by all the secondary organizations in Gaza, and not by the Hamas, or even the Islamic Jihad. As long as Iron Dome continues to intercept rockets, and there are no casualties, the current volley of rockets can go on for a while. If, however, one of the missiles causes serious casualties on our side, the daily tit-for-tat volleys now taking place in the evenings, (after the day long Ramadan fast) may escalate.

Tonight there was very large rally in Tel Aviv's Rabin Square on behalf of the kidnapped students. My best guess is that there were 20,000 people in attendance – maybe even a little more. While the organizers touted the rally as a unifying rally, where secular and religious, settlers and Leftists, could come together, hear some music and reflect – the gathered crowd was overwhelmingly people affiliated with the National Religious. Of course, there were large numbers of non-religious people in attendance. Though they were clearly a minority.

The organizers talked about us being one nation, not really sure how much that is the case. However, as long as there is some hope, regardless of how slim, that the boys may be found alive, we all need to work together towards achieving that end. Sadly, as the time goes on, that hope is getting increasingly more slim. I am happy to say that the

President-Elect Ruby Rivlin spoke briefly about how we have to end all of the killing going on around us, in Syria, Iraq, Sudan and more. I thought that that was an important point to make, but I doubt very many people were listening.

Prime Minister Netanyahu spoke tonight at a seminar for National Security. The bulk of his remarks were directed at the need – especially now that we see what is happening in the rest of the Middle East – to make sure the Jordan River remains our security border. He stated that radical Islam will fail in the long run, but we must be very vigilant in the short term. In that I agree. Stalinism, Maoism and Nazism all failed in the long run, but killed millions along the way. Netanyahu also made headlines by calling for the establishment of an independent Kurdish State. This was the first time an Israeli leader has publicly called for this outcome.

There were two developments today in the ongoing tension between Religion and State. In the first, with the most important long-term impact, the Supreme Court stated that the rabbinic courts could not order anyone to have their child undergo circumcision. The dispute came up a few months ago, when in the midst of a divorce case; it was revealed that the couple's baby boy had not been circumcised. The court ordered the mother to do so, or face daily fines. The High Court ruled that the rabbinic courts had exceeded their authority regarding this matter. I think they exceed their moral authority every time they meet, but that's another matter.

Finally, the Interior Minister used a power I did not know he had to strike down a law passed by the Tel Aviv City Council, officially allowing convenience stores to stay open over Shabbat – something that has been happening de facto for years. I do not shop on Shabbat, but I do shop until a minute before Shabbat. Furthermore, I know many of my fellow Tel Aviv residents who appreciate the convenience of being able to buy some essentials on Shabbat. I have no doubt that Likud Minister Gideon Saar (who considers himself a potential successor of Netanyahu) made this move in an attempt to gain favor with the religious. However, he is doing so against the expressed will of the people of Tel Aviv who voted through their elected representatives to allow the stores to be open.

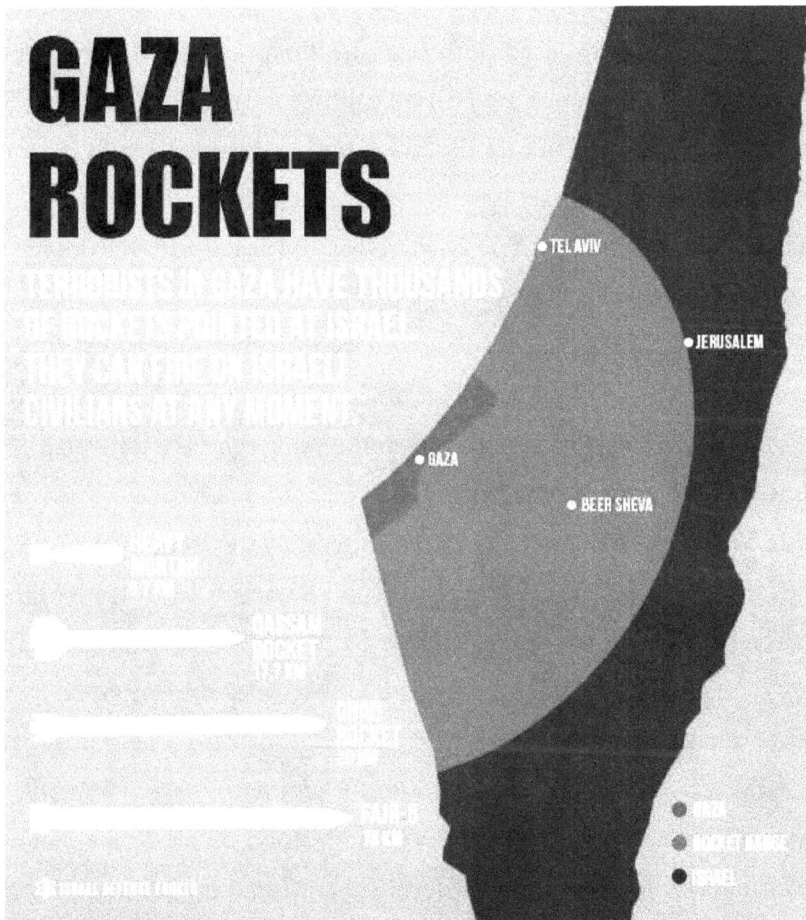

▲Three Gaza rockets struck Sdot Negev on Monday morning, damaging a home in the area. Less than two hours earlier, eight rockets hit populated areas throughout the south.

July 3, 2014
Heavy Rocket Fire on the South, Rioting in East Jerusalem and Religious Coercion in Tel Aviv

There have been so many serious events taking place this past week that it's hard to know where to start. Chanan Kristal, the political correspondent of *Kol Yisrael* summarized it best this morning on the radio: "We have been pushed back 10 years in the West Bank, to the time of the intifada (based both on the kidnappings, and even more so regarding the riots in East Jerusalem). In the area around Gaza, we are back to the period before the last round of hostilities with Hamas nearly a year and half ago. Last, but not least, Minister of the Interior Gideon Saar has put Tel Aviv back at least 20 years, with his decision to overrule the government of the city of Tel Aviv."

Tonight I was faced with the dilemmas of choosing between two important concurrent events: There was a rally calling for an end of the violence and against retribution, at the same time as a long planned teach-in on the increasing religious intolerance and coercion on the entire Israeli population. I decided to attend to the later, in the hopes that (at least in this area) I could have some impact. In fact, I had the chance to express my displeasure to the Deputy Mayor at how Tel Aviv is planning to enforce the old Shabbat rules (but more on that later.) First, let's address the most pressing problem – i.e. the storm of rockets raining down on the South. What started as a few missiles, launched here and there, has turned into a flood of rockets, some of which have unfortunately hit S'derot. Iron Dome cannot really work effectively at that short a range. This leaves the government very few choices, but to act. The Netanyahu government has made it clear that if the rockets stop, it will not take action – However, if the missiles continue we will have no choice but to act. It is not clear what Hamas will decide. There is one school of thought that says Hamas is currently in such bad shape that they have nothing to lose.

▲ (L) *"This city is impossible to stop"* – **Supermarket banners protesting mandated Shabbat market closures in Tel Aviv. (R)** *"It's unpleasant to see a closed am:pm".* These slogans are parodies of popular Israeli song lyrics.

I fear that may be the case, we will see in the next 48 hours or so. The other school of thought feels that because of its weakness, Hamas will not want to take the chance of a major confrontation with Israel. As to events in the Jerusalem ... This week we have seen some of the worst riots we have seen in East Jerusalem in many years. The tragic murder of the Arab teenager has caused a great deal uncertainty, and seems to have clearly undermined our moral position. Unfortunately, it has been nearly two days and there is still no definitive word on whether the teen was killed by Jews or whether he was killed based on some type of family feud. I am not sure what is taking the police so long to make the determination. Once again, this undermines our already limited faith in the police. There is much I have to say about the killings of the young

Yeshiva students, the actions of the army, the government and much more. However, I will wait until after the Shiva period is over.

The one thing I cannot wait to say is how utterly ridiculous all the articles that in some way blame President Obama have been. I have many complaints about President Obama's Foreign Policy choices. Barack Obama is President of the US and not President of Israel. This tragedy was not and should not have been central to his concerns. Finally, back to the item I began with ...

Tonight I attended a long-planned two-part panel discussion (Part II will take place tomorrow) on religious coercion. The planned program was overshadowed, somewhat, by the actions of Interior Minister Gideon Saar to overturn the regulations passed by the Tel Aviv Municipal Council, allowing small convenience stores to remain open on Shabbat. The action has been met by widespread alarm among residents of Tel Aviv – on two levels: First, on the practical level, that most residents want the stores (similar to 7 Elevens) open 24/7 for their convenience. On a second level, it seems far beyond the scope of authority (and reason) for a Minister of Interior from the Likud (and not from a religious party) to dare overturn the sovereign decision of a municipal government – just because he wants to.

Saar's proclamation came as a tremendous shock. I was surprised to hear that the city government intends to fine every store that chooses to open. There were many cries from the floor – Why doesn't the municipality just ignore the ruling, or have a 'blue flu' among the inspectors? The answer was far from comforting. The Municipality asked their legal advisors how to proceed, and were told they had no choice but to heed the Minister of the Interior. Of course, asking your legal advisors in a case like this is what we call in Hebrew a

"She'eilat Kit Bag" (army reference: When a soldier is naive enough to ask their commander if they need to carry their full 'Kit Bag' – with all their gear – One quickly learns not to ask questions to which you really do not want to know the answer.)

The rest of the conference was dedicated to the larger question of the effect of the increased imposition of religious norms on the rest of us – norms that continue to get more and more extreme. A full discussion of this matter will wait for a longer piece.

To my readers in the U.S. – Happy Fourth of July.

A Letter to The Prime Minister
July 6, 2014

Dear Prime Minister Netanyahu,

You are on the road to being the longest serving Prime Minister in Israel's history – and by all accounts, you have made very few mistakes in this position. You have not involved us in unnecessary wars, you have not made any terrible diplomatic blunders, and in some areas you have achieved true accomplishments. Yes, there are many areas in which I could criticize you. There are many policies of yours with which I may disagree. However, that is not why I am writing to you today.

One of your excellent qualities is your ability to communicate – and certainly in English you have no rival. Not since Abba Eban has there been a leading Israeli with such an excellent command of English. Your Hebrew rhetorical skills are no less impressive. Yet you refuse to truly communicate with the Israeli public. Other than your brief appearances with Defense Minister Ya'alon and Chief of Staff Gantz during the hunt for the "kidnapped" boys (who you had to know were dead), you have been largely silent. You delivered an excellent eulogy, but then what?

When the ugly words of revenge were heard you have been largely silent. When the body of the Palestinian teen was discovered (who was most likely killed in revenge), we in Israel learned you were appalled through the Americans. When Arabs in East Jerusalem and in the Galilee began rioting you Left it to others to level threats, rather than speaking out to calm the waters.

Mr. Prime Minister, the South is being attacked again by rockets. So far you have shown restraint, though that restraint may be wearing out. We, your public do not know what you are thinking. Instead of going on TV and explaining what our choices are, and that if we attack Hamas we must be ready for a ground assault on Gaza (something we may be reluctantly forced to do), you still remain silent.

The world around us has been changing radically over the course of the past few months. To our great disappointment, our Prime Minister – a Prime Minister who is one of the best communicators in our history – appears mute. It has been over 700 days since you took questions from the Israeli press at a press conference. Mr. Prime Minister, you lead a strong nation. But your population is confused and worried.

I did not vote for you Mr. Prime Minister, however, you are the only Prime Minister I have. So, please lead. Please speak. Please use your words to calm the waters of hatred. Please use your words to explain our geo-strategic position, and Mr. Prime Minister, please use your words to prepare the Israeli public for a potential military confrontation with Hamas.

As we say to our children, use your words Mr. Prime Minister. And only then be ready to use the other powers that are at your disposal.

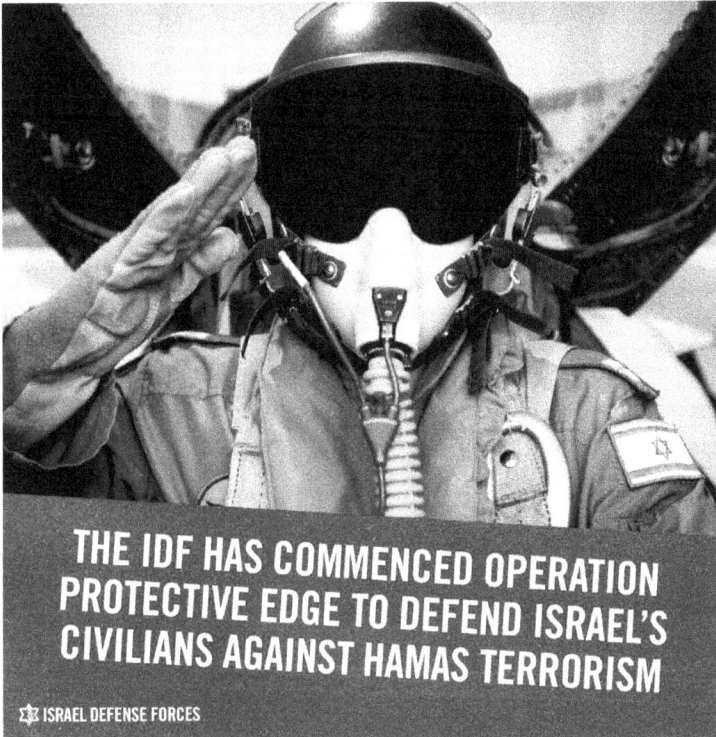

THE IDF HAS COMMENCED OPERATION PROTECTIVE EDGE TO DEFEND ISRAEL'S CIVILIANS AGAINST HAMAS TERRORISM

✡ ISRAEL DEFENSE FORCES

War Begins:
Tel Aviv Diary

July 7, 2014

60 Rockets Fired on Israel in One Hour – All Intercepted

It was a surreal evening. We were sitting in a very nice restaurant with friends visiting from the U.S.; however, it was hard to pay full attention to the dinner conversation. On the wall across from us was a large screen television. Although the volume was off, it was easy to read the titles and much of the closed captioning on the screen. As we moved from course to course, more and more cities in Israel came under attack – from

Ashdod to Be'ersheva, and dozens of places in between. Over the course of our delicious dinner, a total of 70 rockets were fired at Israel. As terrible as that sounds, the only casualties that resulted were three people who tripped while running to a shelter. Why were there no casualties? Because Israel's Iron Dome system, (the anti-missile system that so many said could never work) successfully shot down every one of the missiles that were on a trajectory to cause physical damage or harm human life. It has become almost expected, and therefore, not properly appreciated, but tonight the Iron Dome successfully intercepted 60 missiles in one hour and successfully defended everyone that was in danger.

It does not stop. The missiles continue to rain down on Israel and the world urges restraint. The government has handled the matter with self-control, and the fact that no one has been hurt allows the government to continue to show the restraint, but this can't go on forever ever. Hamas, despite earlier assessments that it did not want a major confrontation, now seems intent on escalating the conflict. What could be their reasoning? Hamas' current situation is so bad they feel that if they are able to get Israel to attack they will regain the sympathy of the Arab world. I think they are wrong.

Things have changed. Other Arabs have killed too many fellow Arabs in the last two years, and I am not sure if anyone is going to care. I hope they pull back from the brink, since at some point we may have to do what we do not want to do. But I am not optimistic.

July 8, 2014
Our Choice in Gaza

Israel faces a difficult challenge in the coming days. In my opinion, Israel has only two choices. It can launch a major ground offensive and recapture the Gaza Strip. Alternatively, Israel can double down on defense and reach the point where we can ignore the continual attacks. Any less definitive action plan will not work.

Our current policy of attacking Gaza from the air will be no more successful this time than it has been previously. There is no way to take action against Hamas that will be strong enough to create the level of deterrence we need against that organization. More specifically, Hamas is counting on Israeli air strikes. They are hoping that despite our care in planning and implementing surgical strikes we will accidentally kill a large number of civilians, and thus, impact world public opinion in their favor. Hamas hopes to regain the support they lost in the past weeks and months on the heels of a tragic Israeli error. Hamas has made the calculated decision to create a crisis. They want us to bomb them just enough to gain them sympathy – Sadly, anything less than that will not create any level of deterrence. It is very easy to fall into their trap.

If we cannot just bomb them, what can we do? One solution is to initiate a ground assault on the Gaza Strip. Such an assault would be similar to the mission implemented in the West Bank after the Pesach bombing in Netanya. It would require a full-fledged assault by all of our active enlisted infantry troops to re-conquer the Strip. This would no doubt result in casualties among our men, as well as to civilians caught in the crossfire. The advantages would be great. We could capture and kill most of the leadership of Hamas and the Islamic Jihad. We could free the Strip from Hamas rule and we could certainly end the rocket fire.

However, then the question remains – What next? I can tell you from personal experience that occupying Gaza (as long as 30 + years ago) was not easy. I can't imagine it would be any less difficult now. Certainly occupying another million Palestinians would not make our lives any easier, on any front.

There is another scenario that might work – That is, if together with recapturing Gaza we reach a comprehensive peace agreement with the Palestinian Authority, that would include them taking over the Strip. I doubt, however, that the members of our current government would be willing to make the concessions that this scenario would entail.

Conversely, we could take a completely opposite approach. In the first Op-ed piece I wrote for the *Times of Israel* in March 2012, entitled: Time for the IDF to Step Up Its Defensive Game. In it, I called on Israel to increase its investment in defensive systems. Clearly that has been done to an extent. However, our natural inclination – honed by years of operational strategy – is for the IDF to act primarily as an *offensive force*. The Air Force hates spending money on defensive systems. The IAF much prefers purchasing overpriced F-35 planes.

Our generals have been trained to attack. Yet, what if we were to double down on defense? We could begin deploying the Israeli developed laser anti-rocket system and increase our Iron Dome capabilities even further. We can reach the point where all of the of missile attacks from Gaza will be intercepted. Then we could be in a position to stop announcing *Tzeva Adom ("red alert")*.

In addition, if they do fire on us, whenever we have the opportunity to respond without collateral damage we can kill their leaders. If Hamas attacks and never does any damage- who cares? There are those who say this is asymmetric - their missiles are cheap and ours are expensive. Though given that our GDP is more than 40 times larger than theirs, we can afford it. To look at the cost-effectiveness of this option in perspective, would anybody like to hazard a guess what an all out assault on Gaza would cost?

We need to start viewing at the attacks from Gaza as a lion looks at a swarm of flies – they are extremely bothersome, but not dangerous. If we put the proper technology in place to make their attacks totally ineffective, (much of which we have begun to do), then we can begin to ignore Hamas instead of repeatedly attacking them. Nothing would undermine them more!

July 9, 2014
DAY 2 @ WAR WITH HAMAS

We are coming to the end of the second day of our "mini-war" with Hamas in Gaza. This war effectively began on Monday night when Hamas launched attacks on many Israeli cities south of Tel Aviv. Once that happened, it was clear Israel would respond with heavy air attacks on Hamas targets, and thus the fighting was on. It has been a strange two days. Last night, when the sirens went off for the first time, we all headed to the bomb shelter in our building. We found that only the entry light worked in the shelter, but that was enough. After two minutes we heard the explosion that signaled the missile had been intercepted. We then headed back upstairs to an evening glued to the TV, with many attacks on different parts of Israel – but no more rockets were sent to Tel Aviv. There was one missile over *Givatayim* (the next town over), which was erroneously announced on TV as an attack on Tel Aviv. There were no warning sirens for this in our neighborhood since the system is designed to only send out an alert if there is a possibility of an attack exactly where you are. However, from our apartment we could hear the explosion when that missile was once again intercepted. Yesterday, during the course of the evening, the Hamas attacks spread as far north as *Hadera*, and also hit the Jerusalem area.

This morning we were greeted with sirens soon after getting up. My son and one a close family friend who is staying with us were asleep. I decided not to try to wake them. Since I realized the night before that with a 40 story building across the street from us (immediately south of our location) there was almost no physical way for a missile coming from Gaza to strike us, even if Iron Dome failed to shoot down the missile.

During the day today Hamas continued to attack (mostly immediately around Gaza). Although they did succeed in shooting one rocket that reached as far as the Carmel Hills area, to the south of Haifa. Tonight they also tried to attack the nuclear facility at *Dimona*. However, all the rockets that might have hit that location were intercepted. Thankfully, in all the attacks to date, not even one Israeli has been wounded The Israeli anti-missile system that is primarily – but not exclusively Iron Dome – has hit 97.5% of the missiles launched that would have hit populated areas. The few misses have been very close to Gaza, where the missiles have very limited time to engage and intercept. Of course, that has not been the case in Gaza, where our attacks have killed 50 people. That is sad, but I am afraid unavoidable.

Hamas is getting more and more frustrated. They attempted to sneak terrorists into the country three times in the last 24 hours. In each case, all of the attackers were killed. Israeli observers are divided on whether Hamas is close to breaking.

Where this goes from here is not clear. There is a lot of talk of a ground assault. By tomorrow night we will have enough troops on the Gaza border to begin a limited assault, but it's not clear if we will actually do so. Before Iron Dome the cost-benefit analysis was clear. We had to stop the rockets at whatever the cost. However, what about now? At this point we are incurring no casualties. Once we send in troops we will suffer casualties. Is it worth it? Do we have a choice? As of this writing it's been a remarkably quiet night. That may be because at the exact moment Hamas was planning to launch its missiles the Israeli Air Force initiated a massive assault on Hamas launch sites. At 11:30 PM Hamas launched 40 rockets at targets in Southern Israel. Iron Dome intercepted seven of them and the rest landed in open fields.

July 10, 2014
DAY 3 @ WAR WITH HAMAS

The third day of our "mini-war" with Hamas is coming to an end. My day began a few minutes before 8 am, when the sirens went off in Tel Aviv. We have stopped going down to the shelter, but stand in a part of the hallway of our apartment that has no windows. This morning, about a minute after the siren, we heard a very loud explosion that shook the apartment. It was (of course) Iron Dome intercepting the rocket in the sky about a mile from us. The missile interception was followed by three weaker explosions (the second intercept and the extra intercept missiles self-destructing.) Two hours later the scenario was repeated. This time the explosions were slightly more muted, meaning the intercepts were probably further south. That was it today, for us. Although there were a few more rockets fired in our general direction, the alarms did not go off (meaning they would not have landed near us), and Iron Dome did its job intercepting the incoming missiles.

▲ IDF
Infographic

Later in the day Hamas fired five missiles at Jerusalem. Two missiles were seen being intercepted on live TV and three fell in open areas – one of them outside an Arab Village in the West Bank. This evening there were two potentially more dangerous attacks. The first one was around dusk, when Hamas launched 50 rockets at Be'ersheva. One of the rockets got through, nine were intercepted, and the rest fell in the desert. Luckily, the missile that got through fell into the yard of a family who were in a shelter. As a result no one was hurt.

Later tonight, a large number of rockets were fired at Ashdod. One missile got through there as well, setting a car on fire. Again, no one was hurt. However, two soldiers were wounded this evening when Hamas fired mortars at them.

Meanwhile, the Israel Air Force continues to attack targets on the Gaza Strip. The IAF claims to still have many more targets they can hit – but the Air Force always makes that claim. The big question hanging over Israel tonight is whether or not to send ground troops into Gaza. In a poll conducted tonight, 47% of Israelis said they opposed doing a ground initiative. On the other hand, there may be no other way to stop the rockets. At the moment, the world seems not to care – with our usual critics remaining rather quiet for the meantime. I could make predictions as to whether we will be going into Gaza by ground, or for that matter, how long such an operation would last. Though quite frankly, any prediction I might make will be no more accurate than using an "Ouija board".

I can say the following – On one hand, the Israeli public is being remarkably resilient in dealing with the current situation. Having Iron Dome certainly takes away the feeling of real physical danger. On the other hand, there is a clear psychological and economic cost. I can't imagine what it was like to live through the London Blitz or equivalent events. I was here for the First Gulf War – when we had no anti-missile defense system. There is absolutely no question that it seems much less dangerous here now. At the same time, it's hard to try to go about daily life, and work productively when a siren can go off at any moment. I can't imagine what life is like for those closer to Gaza, where instead of being subjected to red alerts two or three times a day, sirens are going off every hour.

▲ A kindergarten in central Israel during a rocket alarm

July 11, 2014
DAY 4 @ WAR WITH HAMAS

The Fourth Day of Israel's war with Hamas has come to and end and I believe that we are getting closer to a ground invasion of Gaza. As I wrote before this even began, there were two options: Israel ignores Hamas or plans on an invasion. Unfortunately, Israel decided to do an air campaign first in the mistaken hope that it would work. It is amazing how we continue to make the same mistake by thinking that the Israeli Air Force alone can solve the conundrum. Air forces have always believed they can solve the problem. If we look at World War II, there are two prime examples of failed Air Force attempts to solve military problems: Germany thought their attack on Britain would bring the country to its knees and then the Allied Air Forces thought the strategic bombing of Germany was the solution. In both cases they were ineffectual. The 2006 Lebanon War taght us that the IAF alone couldn't do the job. Thankfully for us, the Iron Dome has made waiting for a permanent solution bearable. I am not a big believer in a ground assault, but I doubt we have a choice at this time. Day Four began with a missile managing to penetrate Iron Dome, landing in a gas station in Ashdod. The result was a fire, much smoke, and thankfully, only one injury. This morning I successfully managed to get to my Friday morning shopping before the sirens went off at about 11:00. The Iron Dome successfully intercepted the missile, however, one of four missiles must have been heading for Herzliya since it was intercepted over Tel Aviv creating a very large explosion overhead and spewing debris on Dizengoff Street.

The rest of the day was quiet and I, and most residents of Tel Aviv, continued regular Friday errands. Unsurprisingly, Dizengoff

Center, which has a food fair every Thursday and Friday, was full of people. Hamas continued to fire rockets primarily on the South throughout the day. In the evening they resumed their attacks on the Tel Aviv area with one attack finding us still sitting around the dinner table. That attack, made up of seven missiles, was successfully intercepted. An hour later, without the sirens going off, but with the knowledge that there was an attack on other parts of the center of the country, another large explosion shook our building, as it seems that again a rocket was exploded overhead. Another day has come to an end. Over 170 rockets have been fired at Israel. Luckily, through it all, only two people have been injured. However, there is no sign of this ending and instead, there are more and more indications that a ground assault may come soon.

July 12, 2014
DAY 5 @ WAR WITH HAMAS
Day Five has come to and end ... My normal Saturday morning routine of bicycling to the beach and walking along the shore was interrupted. Since there are no secure structures along the beach, I decided to stay home. Though by noon it seemed I had made the wrong decision. The morning and most of the day remained quiet. It was only late in the afternoon that Hamas resumed firing on the center of the country – first by firing on the areas around Rechovot and Ashdod, followed by a round of missiles fired at Jerusalem. The Jerusalem salvo caused sirens to go off in the city. However, those rockets all missed, with the exception of one missile hitting an Arab home in Hebron.

At 8 o'clock in the evening Hamas announced it was going to launch new powerful missiles at Tel Aviv. I have to admit that I did take the threat at least somewhat seriously. We made sure when we went to pick up our pizza, (part of our standard Saturday night routine), that we got back before 9 pm. A few minutes after 9 o'clock Hamas kept its promise, and sure enough the sirens went off.

This time we actually went down to the shelter, just in case there was a more substantial attack. After we returned to our apartment the sirens went off again. Back to the shelter we went. Thankfully, as in the first case, the Iron Dome system successfully intercepted the missiles. So, Hamas had made a big deal threatening Tel Aviv and the end result was forcing us to spend ten minutes in our shelter. That's not to imply the situation is not affecting our lives. Tonight we were supposed to go out with some friends, each of us have 13-14 year olds at home. Instead, their family came over and we rented a movie on iTunes.

Hamas has managed to disrupt our lives, to a degree – but only to a small degree. The cafes are still busy. We continue to work, and our lives go on. This is a strange war. We watch TV and watch for messages on the screen announcing where the missiles are headed. Then we watch on TV as they are shot down. When not watching our large screen, my ever-present iPhone app goes off every time a missile is shot. After quickly checking that it's not heading here, I go back to work. Iron Dome keeps working fabulously. Only three people have been wounded on our side during this war. Of course, on the Palestinian side, the death toll has exceeded 100. The war is extremely asymmetrical. A terrorist government that possesses Word War II technology is facing off against a 21st century nation, with the most advanced military technology on the planet. It's clear at this point the only goal Hamas seems to have is getting us to kill their people, something we are reluctantly doing. We have a choice – we can keep killing Palestinians in Gaza (including innocent civilians) and gain some publicity photos for Hamas, or we can send in our soldiers (some of whom will no doubt be wounded or killed.) It's a terribly "Hobbesian" choice. Most other countries would not think twice. We keep rethinking the question…

In two or three years we might reach the point where our defensive systems are impenetrable. We are advancing much faster than our adversaries. However, can we allow our lives to be disrupted every time it suits our enemies? Of course some people say – just make peace with them. I certainly would be willing to do so, (even if it meant giving up all of the lands we captured in 1967), but clearly that would not satisfy Hamas. They still seek our utter destruction, and that is something we will not give them. In the next few days we will see what choices the government makes. There are already rumors of attempts to reach a ceasefire. Though I would be surprised if they bear fruit in the next few days, a ground assault is more likely.

July 13, 2014
DAY 6 @ WAR WITH HAMAS

The sixth day of our "mini-war" with Hamas has ended. Once again, the day began quietly. For the second day in a row there were no missile launches aimed at Tel Aviv this morning. However, in the late morning Hamas achieved one victory. In Ashdod, a 16-year-old boy was walking home from getting his haircut. When the sirens sounded there was no secure space for him to go. He crouched near a wall. Sadly, that turned out not to be enough. A missile landed near him and he was gravely wounded. As of tonight his condition remains critical, but stable.

At 4 pm today sirens broke the quiet of our afternoon, when two rockets were fired at Tel Aviv. Additional missiles were fired at Rishon LeTzion, as well as two longer-range missiles fired at Haifa.

One missile landed – harmlessly – near Hadera and one overshot Haifa. At 8pm we heard a muffled explosion in the distance. After listening to the TV news we discovered that Hamas had in fact fired missiles at Tel Aviv. These missiles missed by such a wide margin, the sirens were not activated, nor were any intercept missiles fired. At that same hour Hamas fired 40 or so missiles at Ashdod and Ashkelon. All were intercepted.

As the days go on we fall into a routine. Regular work continues. We go out of the house to do chores – though we leave our houses with a certain sense of vigilance. We walk down the street always checking where we will go if the sirens go off. Even late at night when I take make nightly bicycle ride, despite the fact I am usually pretty certain that it's too late for them to fire, I still keep my eyes on where I can go if the sirens sound.

As I write this piece (a little after midnight) Hamas has promised a barrage of missiles at the end of the World Cup Finals. They fired about 10 missiles at Ashkelon and Ashdod, when the game went into overtime. I decided to wait a little longer before going out on my ride. Well, I waited a half an hour after the end of the game. There were no rockets from Gaza. However, there was the troubling firing of two more rockets from Lebanon into Israel – both landed near, but not in, the town of Shlomi, in the Western Galilee.

Today there was more talk of a ceasefire. There may indeed be one nearing. Clearly neither side is gaining anything at this point. As I wrote before we started retaliating ... We should either be willing to go all the way and recapture Gaza with ground troops, or ignore them, as a lion does a gnat. Instead, once again, we took the middle road that achieves nothing.

July 14, 2014

DAY 7 @ WAR WITH HAMAS
Ceasefire

I was going to write about the Bedouin girls hurt in a rocket attack, or about the attempt by Hamas to use a drone to attack Israel (that we shot down by missile), but, happily there is better news: It looks like we are about to enter into a ceasefire with Hamas. The Egyptians have called for a ceasefire starting tomorrow morning at 9:00 a.m. Our security cabinet will be getting an early start on the morning meeting (at 7 a.m.), when they will no doubt accept the ceasefire. The Egyptians stepped in after the military arm of Hamas – who rejected earlier requests, turned to them and asked for the ceasefire.

The terms of the ceasefire are for all sides to stop any aggressive actions against the other. Then within 48 hours the sides are set to meet in Egypt separately with the Egyptians to work out additional terms. It's not totally clear what those terms will be, but they will most likely include a return to the status of "quiet for quiet" that existed before the latest flare-up.

As I said in my article this afternoon – calling on our side to announce a ceasefire – we have won this round and it is time to let the country return to normal. There is not a chance we are going to turn down an Egyptian call, especially one backed by the US.

Unless we are willing to risk losing a significant number of men, and know what we do the day after we reconquer Gaza, we must accept the best we can get – and in the meantime even further strengthen our defense systems, (as I have said on numerous times.) Our problems from Gaza are the more manageable ones. Tonight we had additional fire both from Lebanon and Syria. Nothing good is going to come from that. But they will die down for the moment now that there will be a ceasefire.

For anyone thinking of visiting Israel, there are some great deals on hotels and airfares right now!

July 15, 2014
DAY 8 @ WAR WITH HAMAS

The day started with an air of optimism and there was a sense, at last, that this war was over. Maybe the problems had not been solved, but at least the missiles would end and we would have not have lost any people. I was out early with my son for breakfast, and, for a few minutes, we did not worry that the sirens would go off. At 9:00, the Israeli government officially accepted Egypt's ceasefire call. Yet, a few minutes later missiles were flying, beginning with the areas immediately around Gaza. A few minutes later, a major barrage of missiles was aimed at Ashdod. One missile fell on a house, but thankfully the residents were in their reinforced room and were not hurt. Then, a few minutes later, sirens went off in the center of the country and a missile was intercepted over Rishon Le'Tzion. If we had any hope of a potential ceasefire, it quickly evaporated as soon as sirens began to sound in the North, when Hamas launched missiles North towards Haifa.

What happened? Beginning last night, almost all of the experts had said that there was going to be a ceasefire. Overnight, hardly any missiles had fallen. What changed?

First, we had all assumed that the Hamas was not going to say no to the Egyptian President Sisi once again. Second, we assumed that the Hamas was ready for a ceasefire.

What we did not understand was that first, Hamas, at this point, does not care what Sisi thinks. We further did not understand that Hamas was not willing to end the war without achieving anything. The

Egyptian plan did not provide Hamas with any gains at all after a week of fighting and not killing any Israelis. Finally, Hamas has concluded that Israel is weak and not willing to launch a ground assault. Under those conditions, Hamas made the conscious decision to reject the ceasefire and resume the missile attacks.

It is now clear that we are not going to have a ceasefire any time soon. Hamas has only one of two potential goals – either kill us or make sure that we kill so many of their own people that the world will turn against us. I am not sure they are leaving us too many choices.

When I began writing this update, we, in Tel Aviv, had not been attacked yet. Moments after I finished writing, as the piece was being edited, the sirens went off. Three more missiles were headed towards Tel Aviv, and all were demolished by the Iron Dome. At the same time 20+ missiles were sent at other parts of Israel. It looks like it's just another day of living under occasional missile fire.

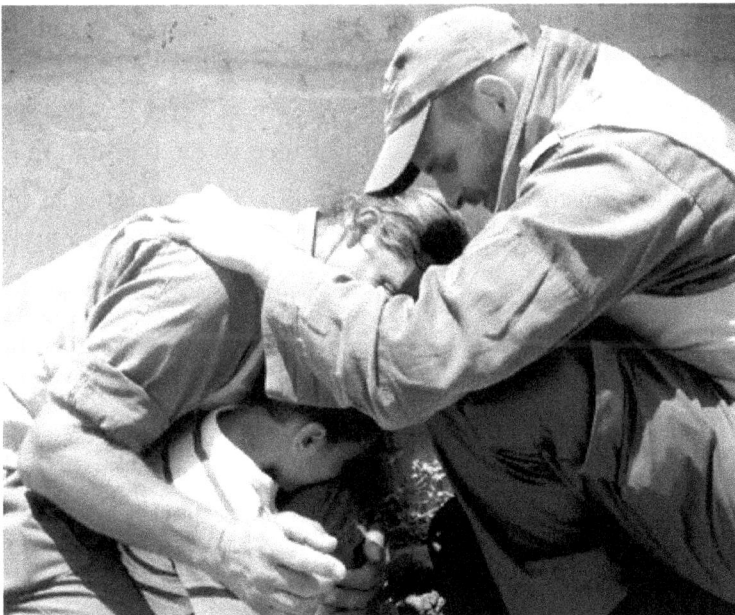

▲ Israeli soldiers shielding a 4-year old boy during a rocket attack.

July 16, 2014
DAY 9 @ WAR WITH HAMAS
The New Normal?

Today is the ninth day of the war between Hamas and Israel. It's the ninth day that missiles were fired at Tel Aviv and the ninth day that all of those missiles were downed by Iron Dome. Other parts of Israel were not as lucky. However, thankfully, today no one was wounded, due to the fact that Israelis (in an uncharacteristic manner) have been following the instructions of the Home Front Command and taking cover in secure places whenever the sirens go off. The residents of Gaza, of course, were not as lucky. While Israel continues to attack the Hamas – in response to their attacks on Israel – Gaza's civilians continue to be wounded or killed. Tonight the news in Israel is dominated by the reports of what seems to be the result of a tragic mistake, in which four children were killed on a beach of Gaza.

Today was the first day since the war began that I ventured far from my home in Tel Aviv. This morning I had a meeting in Jerusalem, a meeting I decided not to delay. As I headed to the bus station (a 20 minute walk from my house), I kept making what, in normal times, would be considered a ridiculous decision – should I walk on the side of the street with shade, but where it would be difficult to find shelter if the sirens go off, or should I just walk in the sun. When I got to the bus station, for the first time in my memory the bus was only half full when it left for Jerusalem. Clearly people are staying inside and not venturing far from home.

My major concern in traveling was not for my safety, but rather, my fear of an attack on Tel Aviv while I was away. Sure enough, as the bus reached the halfway point to Jerusalem (a 50-minute ride), my phone rang. My son was on the phone. The sirens were going off and he, my wife, and the guests staying with us were in the secure location. I stayed on the phone with him until he heard nine explosions (the biggest consecutive number of the war), each explosion representing another successful missile interception. Jerusalem seemed more relaxed than Tel Aviv. Rockets have been fired at Jerusalem, but not nearly as frequently as on Tel Aviv.

When I returned to Tel Aviv in the afternoon I met one of my closest friends for coffee. Next week he is heading off to a quieter place- Kiev – to consult for the government of Ukraine. Sitting drinking coffee in the financial district of Tel Aviv it was hard to believe that multiple missiles had been fired on the city just hours before.

So where are we headed? Tonight Hamas officially rejected the Egyptian ceasefire plan. They have countered with demands that go far beyond what anyone would even consider. There continue to be mediation efforts by the Egyptians. The U.S. had been working with Qatar and Turkey, who Hamas is happy to have involved, but whose involvement both Egypt and Israel categorically oppose.

Tonight there seem to be additional efforts in Cairo to reach an agreement. The Israeli government is still optimistic that a ceasefire can be reached. However, the Israeli government has been consistently wrong about the intentions of Hamas over the past few weeks. Prime Minister Netanyahu is very reluctant to send ground troops into Gaza, despite the fact that his reluctance is costing him support from his traditional right wing supporters. At this point Hamas does not seem to have any intention of agreeing to a ceasefire, and Israel really does not want to invade. Something is going to have to give. The "new normal" cannot continue. The economic and psychological costs on Israel are too high. Israel's current actions have not deterred the Hamas. Tonight the government approved the draft of an additional 8,000 reservists. Does that mean that Israel is about to send in ground troops? It seems unlikely, at least in the next 48 hours.

▲ Hamas' underground tunnel network in Gaza made it possible to infiltrate Israel and ambush civilians and IDF soldiers. A primary objective of Operation Protective Edge was the elimination of this threat.

July 17, 2014
DAY 10 @ WAR WITH HAMAS
Israel Launches Ground Assault

Israel has begun a ground assault on Gaza. This action was coming all day. When Hamas once again refused the Egyptian plan of a ceasefire. Hamas fighters attempted to infiltrate into Israel through a tunnel. The Netanyahu government felt they had no choice. As of 10:15 p.m. Israeli troops began entering Gaza with a limited goal of destroying the tunnels that Hamas has dug into Israel. It will require Israel seizing a two-mile strip of land – a mission, which the Israeli army expects, will take approximately two weeks.

The pressure on Prime Minister Netanyahu to order a ground assault has grown over the course of the past few days. That pressure built from three points. First, Hamas was uninterested in agreeing to a ceasefire. The Egyptians – who are totally frustrated – warned Hamas that they would reap what they sow, (meaning, eventually Israel will go in to Gaza.) Second, infiltration into Israel proper through the Hamas built tunnel renewed the plea by many in the army to be allowed to neutralize that avenue of entry. These are 2-mile tunnels running from inside Gaza into Israel. Today was the second time Hamas tried to use these tunnels to attack Israelis by surprise. An estimated 10 such tunnels exist stretching into Israel from Gaza. Finally, domestic pressure was mounting on Netanyahu and the government, not only from the right flank, but also from a large portion of the public – who while they appreciated Netanyahu's steadiness, felt he was just delaying the inevitable.

Today was day ten of the war, a war that will now last for at least another 14 days. I cannot imagine what it was like to live through the Battle of England. The original version of this article said that we in Tel Aviv had not heard any missile alert sirens today. However, as I was editing this piece, the sirens went off. Again, we rush to the protected area – explosions in the air – another attack on Tel Aviv. Seven missiles fired, seven missiles shot down. This morning I went out with my son for what has become a morning routine – breakfast at a nearby cafe. As we got up to leave the cafe we started receiving news alerts on our phones of a missile attack on the Coastal region. No sirens went off.

We rushed home. Suddenly we heard the relatively distant sound of explosions in the sky. This was the sound of missiles directed at a neighboring town being shot down. We arrived home to find out that although there had been a failed missile attack on our area, the major news item was of a different sort of unsuccessful attack. There had been an attempt by Hamas to sneak heavily armed fighters into Israel, in order to kill and kidnap members of a Kibbutz near the border. News reports were showing images caught on camera by Israeli observation points of 13 terrorists emerging from a tunnel inside Israel. Within a minute of their sighting, an Israeli aircraft fired a rocket at them, causing the terrorists to return to their tunnel and escape. The aborted attack through the tunnels was the final push that convinced the Israeli government to attack.

At 10 a.m., a five-hour humanitarian truce went into effect, as requested by the United Nations. During that time, people in Gaza when out to get basic goods. In Tel Aviv it was a chance to take care of errands that had been put off. For a few hours, even the streets of Tel Aviv were a little more crowded than they have been recently. A few mortars were fired from Gaza, in the middle of the ceasefire, but by and large the ceasefire was observed. At the same time, rumors spread there was going to be a permanent ceasefire starting from 6:00 a.m. tomorrow morning. After reading my Twitter feed I looked up at one of our house guests (who has been largely stranded in our apartment) and said: "hopefully after tomorrow you can start traveling again." However, that was not to be. It soon became clear that it was a false report. Once again, it seems that Israel agreed to an Egyptian proposal, but Hamas did not.

When 3 p.m. arrived, Hamas resumed its attacks on Israel by first firing missiles at Be'ersheba, Ashkelon and Ashdod. An hour later they tried, once again, to attack Tel Aviv. Once again, the sirens did not go off. Since those missiles were not going to land on Tel Aviv, but rather a city next door. This time, since I had not gone to our secure space, I had a bird's-eye view (right out my living room window) of an Iron Dome Tamir rocket intercepting a Hamas missile over the Tel Aviv metropolitan area. So now, I can sit on my living room couch and watch a war in real-time. This is a surreal experience.

A few other interesting developments: This afternoon UNRWA (the United Nations agency that provides relief for Gaza residents) complained they had discovered 20 Hamas missiles in one of its schools. What was unusual was not that the missiles were there, but that UNRWA complained. Tonight Hamas tried, once again, to fly a small, armored remote-control plane into the city of Ashdod – and once again, the plane was intercepted and destroyed by a Patriot missile.

Prime Minister Netanyahu made the announcement tonight that Israel was sending ground troops into Gaza. The goal of the mission is limited, but once you enter, you never know what can happen. The reaction of the average Israeli to the announcement and what they are all seeing live on their television reinforces is that they will need to live through a minimum of another 14 days of rocket fire. There is a grim determination that there is no choice, there are no easy solutions, combined with a hope that as few lives as possible will be lost in the coming days.

July 18, 2014
DAY 11 @ WAR WITH HAMAS (Part 1)
A Just War

War is a terrible thing. It is something to be avoided. It is an action of last resort. For many years we have gone to war a little too easily. We have gone to war to achieve political objectives, which in retrospect were unachievable. This time it's very different. For the first time – since the 1973 Yom Kippur War – we are in a war that was not of our choosing; a war that we did not want to flight; a war that was forced on us by our enemy (an enemy who admittedly does not care what effect the fighting will have on his own people). For the first time in 40 years we are fighting a war which there can be no doubt is a *"Milchemet Chova"*, an obligatory war.

We entered into a ground war after 10 days during which Israeli cities were continually and extensively attacked from the air. Prime

Minister Netanyahu tried every way he could not to send in the ground troops. He waited to respond after the first missile attacks. He has responded positively to every possible ceasefire plan. Time and time again, we have seen Hamas respond by firing more missiles. Hamas has made it clear that it only wants one thing – war.

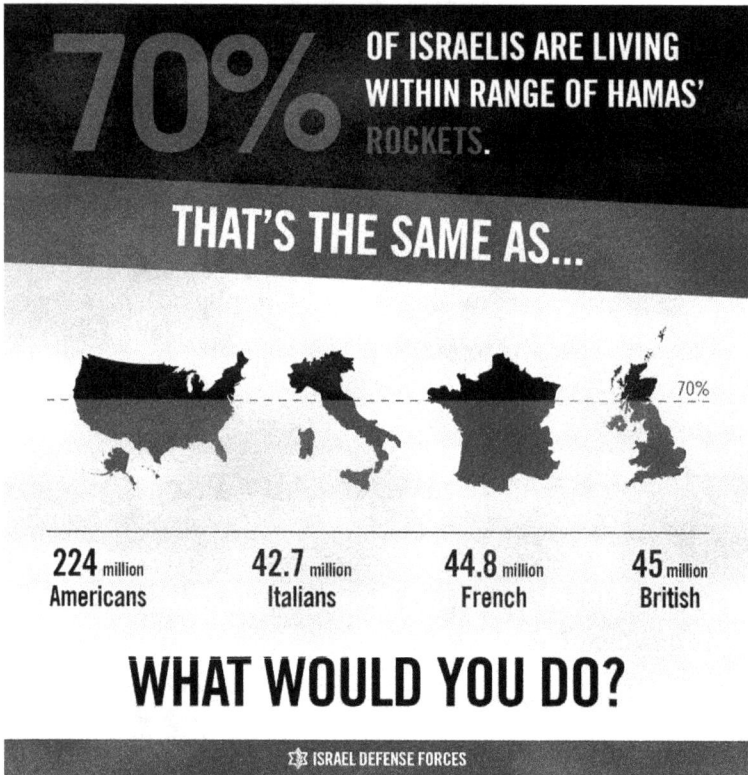

70% OF ISRAELIS ARE LIVING WITHIN RANGE OF HAMAS' ROCKETS.

THAT'S THE SAME AS...

70%

224 million Americans   **42.7** million Italians   **44.8** million French   **45** million British

WHAT WOULD YOU DO?

ISRAEL DEFENSE FORCES

For once the world understands why we have to act. Yes, the news media will no doubt show terrible pictures in the coming days of Palestinian woman and children who are killed. Those deaths are truly tragic. However, those deaths cannot be on us. We did not ask for this war. Many innocent Germans died from the Allied attacks on Germany, some of which could have been avoided – in retrospect. However, that fact did not make the war on Germany any less moral.

For the moment, Israeli politicians – from Right to Left – are all united in support for the government. The only rational argument that can be made against the ground attack is that it possibly could have been avoided – "if we only had reached a peace agreement with the Palestinians". Counter factional history is always possible. However, keeping in mind that Hamas' goals remain our total destruction, it's not a very strong argument.

The world will soon argue that we must remain proportional, or well meaning interlopers (like US Secretary of State Kerry) will urge we limit our actions to dismantling Hamas' tunnels – and maybe we should. Though maybe we should go further. This time it must be our decision. We were forced into this war. Now our sons and, in some cases, our daughters are fighting for the very goal that Zionism came about – to ensure our survival as a free and independent people.

July 18, 2014
DAY 11 @ WAR WITH HAMAS (Part 2)
Day One of the Ground War

Today was the eleventh day of the war between Israel and Hamas; the eleventh day of hearing the sirens go off, of taking shelter, and then hearing the explosions as the missiles are intercepted. Yet, today was somehow different. It was the first day of the ground assault into Gaza. You could feel it in the streets and in conversations with people. People are no longer worried about inconveniences and of the possibility of being endangered by a daily barrage of missiles. Instead, now, some of their children and/or husbands are, or would soon be, engaged in a ground war- a war that, however large our advantage was, would no doubt result in some casualties. Overnight, the number of those called to reserves increased, touching the children of friends. People who had been wondering whether to go on a planned trip, either for business or pleasure, quickly decided this was not the time to be away.

My son woke me up this morning to tell me that the first soldier had been killed. I did not know this young man, but it felt like an intense, sharp punch to the gut. It was only later in the day that I found out he had served in the same unit in which many of the soldiers I know are currently serving. A bit later, we learned that the first fallen soldier was a casualty of a friendly fire accident. I am sure there are those who, reading this, wonder how I can be so concerned about the death of one Israeli soldier when so many civilians, including children, are dying in Gaza. Truth be told, they would be right. We should all care about any person killed. However, inevitably, we all identify more with our own. And, many of those now in Gaza are kids that I know.

This war is very different from other recent wars. Across the Israeli political spectrum, there is an acceptance that this war was forced upon us. This is a war that nobody wanted. This is a war that Israel tried to avoid, having agreed to a variety of various ceasefire proposals put forth. As a result, support for at least a limited incursion into Gaza is nearly universal. However, if the scope of this war expands, that could change.

As to the war itself, by all indications the ground actions have been going well from the Israeli perspective. According to a report, half of the expected 12 tunnels have been discovered and work has begun to destroy them. Casualties have remained very light. Other than the soldier killed from friendly fire, there have been only a few cases of lightly wounded soldiers.

▲ IDF Armoured Brigade Corps near the Gaza border. July20, 2014.

Meanwhile, on the home front, the missiles have continued to fall. The cities surrounding Gaza have been fired on continuously. Sitting in Tel Aviv, as the day was coming to end, it was clear that at least one attack would happen soon. The Hamas has never let a day go by without firing missiles at Tel Aviv. At six this evening, the sirens began sounding. Another attack. Five more rockets fired, and five more rockets intercepted by Iron Dome.

How this war will end remains a big question. Prime Minister Netanyahu warned today, at the opening of a cabinet meeting, that he had ordered the military to prepare for a wider action. The Hamas has indicated that it has no interest in ending the conflict at the moment. It believes it has nothing to lose. Perhaps in a few days, with 30-40,000 Israeli troops already in Gaza threatening to expand their operations, the Hamas will finally take Israeli threats seriously and agree to bring this war to an end. Then, of course, I wrote maybe – but maybe not. And that is what worries many Israelis tonight.

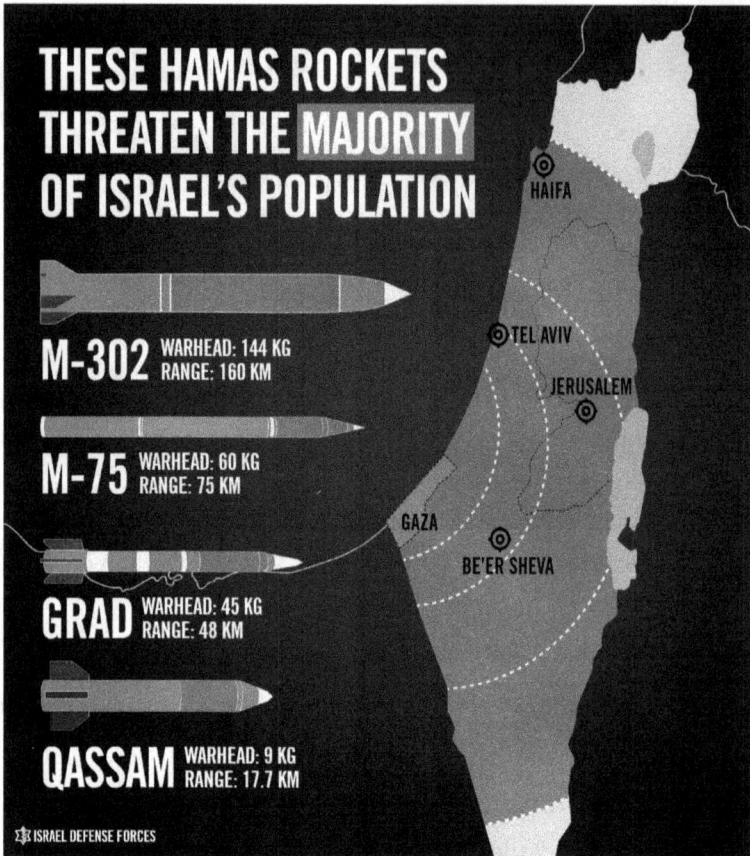

THESE HAMAS ROCKETS THREATEN THE MAJORITY OF ISRAEL'S POPULATION

HAIFA

M-302 WARHEAD: 144 KG RANGE: 160 KM

M-75 WARHEAD: 60 KG RANGE: 75 KM

GRAD WARHEAD: 45 KG RANGE: 48 KM

QASSAM WARHEAD: 9 KG RANGE: 17.7 KM

TEL AVIV
JERUSALEM
GAZA
BE'ER SHEVA

ISRAEL DEFENSE FORCES

July 19, 2014
DAY 12 @ WAR WITH HAMAS

It is day twelve of our war with Hamas and it certainly not looking like a quick war. Last night, we had friends from the United States over for dinner. We had spent ten years convincing them to come visit Israel, assuring them not to worry, as it is really a safe place. They arrived two weeks ago, and left last night. Before leaving, they came back to Tel Aviv just to have dinner with us. Before dinner, they went to the beach and had the opportunity to experience an attempted rocket attack on Tel Aviv. There, they had the chance to crowd into a public safe room alongside a public bathroom. As our dinner ended, and they were set to head for the airport, Hamas decided to give them a sendoff, and the sirens once again went off. We all crowded into our safe zone. When the sirens ended, they headed off for the airport to go back home.

Most Saturday mornings, I bike to an area in North Tel Aviv and spend an hour walking along the quiet beach. That was out of the question today, since there is no place along that beach to take cover. So, instead, I walked with my son to the nearest beach – a 15-minute walk from my home. The beach was relatively empty for a Saturday morning.

I spoke to a few of the people there. There was one group, which consisted of two middle age couples who said they had been coming to the same spot every Saturday morning for 20 years. Nothing would stop them from getting together. They called themselves a parliament, which is local term for a group of people who get together regularly to discuss the situation. As a group, they were very satisfied with the way that the Netanyahu government was conducting itself, even though they were not Netanyahu supporters. One of them stated that he felt the operation had been important in helping solidify Israel's deterrent status, especially with Hezbollah in Lebanon and with Iran, who now would have to be very concerned that their missiles would be intercepted and thus would be less likely to attack. They were all concerned about the economic costs of the operation. And, like everyone, they were concerned for the wellbeing of the troops. They were also concerned with what would be the day after.

I had a second conversation with a British immigrant who has been here for three years. He was at the beach we were at and not the one he usually goes to with his 1-year-old daughter, since this beach had a safe place to run to if the sirens went off. He was more pessimistic than the first group and was very concerned that the government had no long-term plan to make peace. When asked if we could make peace even if we were willing to make all of the needed concessions, he was hesitant, saying we live in a very tough neighborhood.

On the military front it has been a mixed day. Two Israeli soldiers were killed when Hamas fighters used one of the tunnels that run into Israel and emerged to see an army jeep carrying unit commanders. They opened fire on the jeep with an RGP and killed two of the passengers. The soldiers returned fire and the Hamas unit that was planning to attack a nearby Kibbutz returned to Gaza after losing one of their members. In a second incident, two Hamas members emerged from a tunnel inside Israel where an army bulldozer was working. They were run over and killed by the bulldozer. Tranquilizers were found on their bodies, as their plan was to kidnap Israelis. The army activities inside Gaza went well, with additional tunnels found and destroyed and Israeli troops sustaining only a few lightly- wounded troops. Special forces have been operating in other parts of Gaza, trying to attack missile sites. Tragically, Hamas fired a missile towards Israel's nuclear plant in Dimona. But, instead, they hit a Bedouin encampment in the desert and one man was killed, with four other wounded.

There seems to be no diplomatic movement, with Hamas unwilling to reach a ceasefire on any terms that are nearly acceptable to either Israel or Egypt. Egypt, who Israel is familiar with as a good mediator in situations like this, currently see Hamas as a greater enemy than even Israel does and thus is not rushing to work hard as a mediator. With no diplomatic solution on the horizon, and without the Hamas seemingly feeling enough pressure, the sense in Israel is that the government will decide in the next day or two to expand the ground operations in order to put additional pressure on Hamas. At the moment, there is very little optimism that this war will end anytime soon.

July 20, 2014
DAY 13 @ WAR WITH HAMAS
13 Israeli Soldiers Killed

The rumors swirled all day. Something had gone very wrong in Gaza last night. All day long everyone knew that something was wrong. But, no one knew exactly what. Towards evening, the news was released.13 soldiers of the Golani Brigade died last night, the majority in two ambushes when the troops moved into the Shujaiyeh neighborhood, a built up section that is closest to the Israel border. The brigade moved in to find tunnels that were believed to lead back to Israel. When they were ambushed, they responded with intense fire and, as a result, a significant number of civilians were killed. Israel had repeatedly requested that the civilians leave the area. They did not and, according to independent observers, the failure to move was a combination of not knowing where to move and Hamas insisting that they not move. The pictures of civilian deaths coming from the Shujaiyeh area have been terrible and during the day Israel refused to report what had happened, waiting to make sure that all of the loved ones of those killed were notified. Tonight, the story was clarified and it became clear that the deaths were the result of difficult battle.

Prime Minister Netanyahu and Defense Minister Ya'alon found it necessary to address the Israeli nation tonight to express their public condolences to those who died and to ensure the public that the government was doing all it could to protect the soldiers and make sure that Israel would accomplish its goals. Defense Minister Ya'alon stated tonight that there are a number of hard days of fighting ahead and that this war will not end soon.

At the moment, it is not clear how this war will end. Hamas does not seem willing to enter into a ceasefire on terms that are acceptable to anyone else. Prime Minister Netanyahu was asked at tonight's press conference if he has been in touch with Abu Mazen during the confrontation. Netanyahu did not give a direct answer. The answer he did give was very interesting. Netanyahu said, "Israel has been in touch with many different people in the Middle East, including the Palestinian Authority." He stated further "it is clear that we could not reach any sort of agreement with Hamas as it is, and everyone understands that. No one but Iran and Qatar likes Hamas, and Israel now has shared interest with the Palestinian Authority, which opens some interesting new opportunities for diplomatic progress".

There are some other interesting signs. Observers in Gaza report that people in Gaza are finally beginning to express their displeasure for Hamas, and there is a growing realization that all Hamas is doing is bringing more misery on the people of Gaza. Hamas has managed to kill Israeli soldiers in Gaza but it seems to be running out of long-range missiles. In fact, yesterday there were no attacks on Tel Aviv and as of this writing late Sunday evening in Tel Aviv, the sirens have not gone off. It has been 48 hours since the last time the sirens in Tel Aviv have ton off. Tonight, the security cabinet is meeting to decide on the next steps. Opinions are divided and it is not clear what will be required to convince Hamas to agree to a ceasefire. Everything from a unilateral ceasefire to going all the way into Gaza to destroying Hamas seems to be on the table. I do not believe, after today's losses, that Israel will agree to end the war with just a ceasefire, based on quiet for quiet. Tonight, the White House has announced that Secretary of State Kerry is coming to help negotiate a ceasefire. Some Israeli commentators have responded, based on his previous "successes," guaranteeing that there will be no ceasefire. There will be one, eventually. It is only a question of how many will die before that happens.

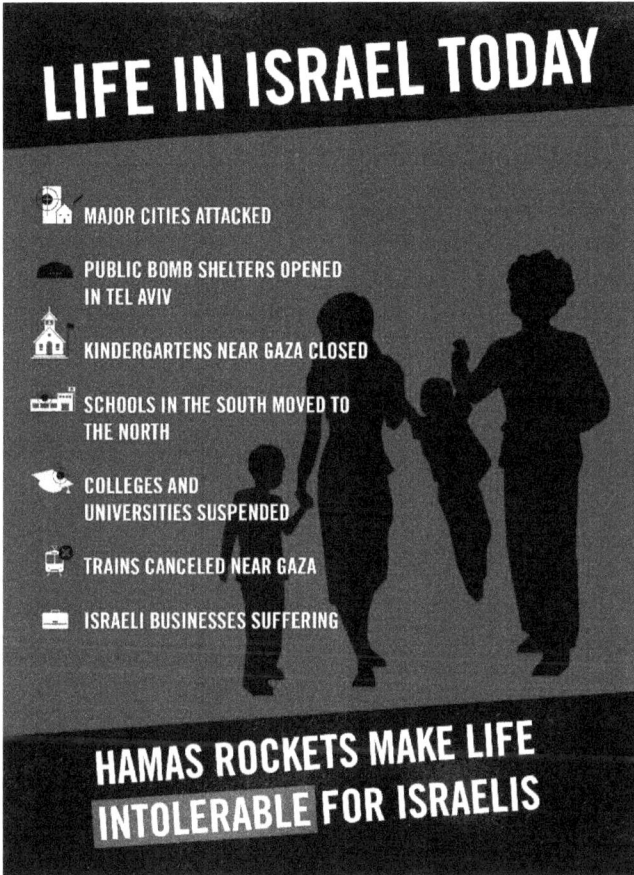

LIFE IN ISRAEL TODAY

- MAJOR CITIES ATTACKED
- PUBLIC BOMB SHELTERS OPENED IN TEL AVIV
- KINDERGARTENS NEAR GAZA CLOSED
- SCHOOLS IN THE SOUTH MOVED TO THE NORTH
- COLLEGES AND UNIVERSITIES SUSPENDED
- TRAINS CANCELED NEAR GAZA
- ISRAELI BUSINESSES SUFFERING

HAMAS ROCKETS MAKE LIFE INTOLERABLE FOR ISRAELIS

July 21, 2014

DAY 14 @ WAR WITH HAMAS (Part 1)

What Liberal Jews Don't Understand About Us

For the past few years, (both before I returned to Israel and the years since), I have had an ongoing disagreement with some of my American Jewish friends about J-Street and other Left-wing Jewish groups that have been critical of Israel. Let me say – from the start – that I stand firmly on the Left of the Israeli political spectrum. I voted for *Meretz* in the last election, and (other than this past week), I have never been a supporter of Prime Minister Netanyahu. However, any lingering sympathy I had for J-Street, or others from the U.S. critical of Israel has dissipated in the past few days.

This morning Allison Kaplan Sommers wrote an article announcing that *J-Street pulls sponsorship from Pro Israel Rally in Boston*. According to Kaplan Sommers, J-Street pulled their support of the rally because there was "no voice for our concerns about the loss of human life on both sides". This article came on the heels of may anger at Peter Beinart last night. I follow Beinart on Twitter, (I follow about 140 people worldwide) and have been frustrated by the fact that the only items he has re-tweeted to date were articles and tweets critical of Israel. Never a tweet from the IDF, nor even from any non-official pro-Israel source.

Now I realize the nature of the vast gulf between my beliefs (together with the majority of the Israeli Left) and the beliefs of American Jewish critics, who claim to be pro-Israel. To us, the current war with Hamas is a just war. We may not support our government all the time. However, when a war begins; a war we did not start and did not want, we support the government and our soldiers.

Yes, I profoundly regret the loss of human life in Gaza, but quite frankly when "children" I know are fighting in Gaza and rockets are being fired on them from homes in which civilians are likely located (civilians who were given many warnings to leave) I prefer that the Air Force send a missile into that building, instead of having one of our boys killed by an RPG fired from that same building. In this case, I feel deep sadness about the innocents that may die, but I know we have no choice.

I watch on TV as Gaza based terrorists emerge from tunnels prepared to attack a Kibbutz and kill whoever is there. I run to shelter when the sirens go off as missiles head towards me. I know Hamas would celebrate if my family and I were killed by an incoming missile. However, that will not happen because we have spent the time and money to develop Iron Dome.

Decisions in this world are usually a spectrum of shades of grey – but this time it is not. The morality of our current conflict is black and white. If J-Street and Peter Beinart do not understand that, I do not need their advice on other matters. I also oppose our settlements in the West Bank and would be willing to go very far to achieve a peace that I have serious doubts is obtainable. However, if our "friends" in the Jewish community hope to have any say about our future, they must learn to understand us in times like this. Otherwise, please keep your judgments to yourselves. Somehow we here in Israel will sort out our future ourselves.

July 21, 2014

DAY 14 @ WAR WITH HAMAS (Part 2)
7 More Israeli Soldiers Killed

Around 11:00 AM this morning, after going over 48 hours of without any rocket attacks, the possible sense of normalcy that had returned was shattered when the sirens began their wail. They seemed to go on for longer than usual. Maybe it was just that we had not heard the warnings for a few days that made them sound longer and louder. It turns out that Hamas had fired 18 missiles simultaneously at points in Israel – two were aimed at Tel Aviv. They were all shot down. Late in the afternoon another cluster of rockets was shot targeting central Israel. All were shot down south of Tel Aviv. This made it unnecessary to set off the sirens in central Tel Aviv. Tonight additional rockets were fired at central Israel.

This day began with another Hamas infiltration into Israel proper via one of their numerous tunnels. This particular tunnel came right up into one of the Kibbutzim in the Gaza area. The Hamas fighters who emerged were dressed to look like IDF forces. As a result, they were able to gain a moment of surprise. In that blink of an eye they managed to kill four Israeli soldiers. The entire Hamas squad of ten was then killed by Israeli fire. This successful infiltration (the 5th such attempt since the beginning of the war) underscored how critical the tunnel threat is to Israel.

The four Israeli soldiers killed in the tunnel ambush were joined by three other Israeli soldiers who died today – two apparently from friendly fire. That brings the number of Israeli military deaths to 25 in the fighting over the past four days. These fatalities are significant compared to only ten soldiers lost during the last Israeli ground incursion into Gaza. In the last 24 hours there has been considerable hand wringing about the way that seven of the casualties died yesterday in an attack on an APC (Armored Personnel Carrier). The APC was nearly 50 years old, and it was the policy of the army not to allow older APCs to enter Gaza.

On the ground, the IDF has found a number of additional tunnels in the last 24 hours- tunnels they were previously unaware existed. It is believed that it will take two to three more days to destroy all of the tunnels they have successfully found. The fear is that not all of the existing tunnels will be found, even during an extended operation. Meanwhile, though the army has faced heavy opposition in a number of points, at most places the opposition has been much lighter. The IDF has concentrated its mission on identifying and dismantling the tunnels, but has expanded its effort to destroy some of the missile launching sites.

On the diplomatic front, U.N. Secretary Ban Ki-moon is in Cairo, while U.S. Secretary of State Kerry is on his way to the region. Kerry seems a little like an uninvited guest, since the Israelis and the Egyptians have shown no great interest in his mediation, (many silently thinking his time could be better spent dealing with the problems in Eastern Ukraine). However, regardless of local sentiments Kerry has arrived in the neighborhood. There is talk that in two or three days there could be a ceasefire. Any ceasefire is still dependent on Hamas being willing to accept it. Opinions in Israel are divided as to whether Hamas is going to feel it has any reason to end the war it began.

July 22, 2014
DAY 15 @ WAR WITH HAMAS
A Frustrating Day

Things in Israel appear to have changed in the last few hours and not for the better. Today started with a morning round of missiles. It seemed routine ... another round of missiles targeted at Tel Aviv and the surrounding area. For us, it was just another attack. We woke our sleeping guests and all squeezed into our secure area. The siren wailed for a few minutes. We heard the loud booms of the Israeli intercept by the Tamir missiles over our heads, and then moved back to the living room to hear what happened. We learned that the missiles aimed at Tel Aviv were downed, but that at least part of one rocket came down in a suburb, about 20 minutes outside of Tel Aviv. We really did not give it much thought. I know that suburb well, having been there many times. It is located close to Ben-Gurion Airport. Thankfully, no one was hurt and the damage was mainly to items stored in the yard.

Late in the afternoon word started spreading through the United States that the FAA had ordered all American airlines to stop flying to Israel. A Delta plane on the way to Tel Aviv landed in Paris and passengers were told they were on their own. U.S. flights from Tel Aviv were canceled, and the planes that were here took off dead head – with no passengers on board. This evening I heard from a friend in Brussels, who was happy to tell me she was leaving to come home in 1/2 hour on Air Brussels. 20 minutes later she messaged me that her flight was cancelled and she was stranded. The Israeli airlines do not have the capacity to replace all the cancelled seats on the other airlines. However, tonight they are talking about creating a temporary air bridge to Cyprus where the foreign airlines can pick up the passengers.

The other items in the news have also not been exactly optimistic. First, as expected, the diplomatic efforts all ran aground due to Hamas' refusal to accept a ceasefire. For reasons that I do not understand, both U.S. Secretary of State Kerry and the U.N. Secretary General came here somehow believing that because they arrived on the scene, an organization (that is at its core) a terrorist organization, would suddenly agree to a ceasefire. It should have been clear to Kerry that the pictures of dead Gazans on the TV that seem to so profoundly upset him and President Obama (as they should) do not move the leadership of Hamas at all. To Hamas, the deaths of their people are of little concern to an organization that practically invented suicide bombing.

The mood in Israel has not been helped by two additional factors – the constant coverage of the funerals of the soldiers who have died. The Israeli media covers every funeral, telling a short life story of each and everyone soldier. Each loss is felt by the whole country, and with that number up to 28, the cumulative effect is strong.

A further complication in the national morale is the story Hamas announced two nights ago – that it had kidnapped an Israeli soldier. The

I.D.F. never quit denying the abduction. Today the whole complicated story came out. The soldier involved was in the APC that was hit by a missile three nights ago. There were seven soldiers in the APC when it was hit. The APC burned for three hours.

When the APC fire was put out, it was hard to identify the bodies. One soldier's body, Oron Shaul, has not been found. The soldier is officially categorized as "missing", but the general consensus is that he is dead. Hamas distributed pictures of kidnapped soldier, but they were all proven to be fake. However, it turns out, as psychological warfare it is very effective.

Two other pieces of news tonight – UNWRA just announced it found missiles – again tonight – in one of its schools being used to house refugees. Last time they just turned the missiles over to Hamas. Finally, tonight, there was a little good news from the Israeli perspective – when the European Foreign Ministers called, unanimously, for Hamas and other armed groups in Gaza to be disarmed.

Israelis are settling in, with grim determination, for a war that is going to continue for a number of days. I expect Israel will try to increase its pressure on Hamas in the next 24 hours ahead.

July 23, 2014

DAY 16 @ WAR WITH HAMAS
Another Frustrating Day

This was a day of reflection in Israel. The number of rockets falling on Israel's cities has lessened in the last few days. While Hamas has continued to send rockets to the center of the country, the numbers of rockets have dropped off by 50%. One more person in Israel, however, died - a Thai worker working in one of the hot houses at a Moshav not far from Gaza was killed by mortar fire.

In the field, three Israeli soldiers were killed when a building they were entering was blown up. Israel destroyed an empty hospital that was used as a command center by the Hamas. The number of tunnels dug to go into Israel has risen to 26, and the sense is that there are more.

Two both heartwarming and simultaneously sad events have occurred in the last two days. Two funerals of American citizens- two young people who decided to come to Israel and fight in the Israeli army- were killed in the current war and were buried this week. In both cases, the two soldiers were considered lone soldiers - soldiers who are here without parents. And, in both cases, the word spread on Social Media for the average Israelis to attend the funerals so that they should not be alone at the time of their internment. 40,000 people showed up in Haifa for the funeral of Sean Carmeli and 30,000 showed up to the funeral of Max Steinberg of Los Angeles today on Mt. Herzl.

Last night, I came across two very interesting groups of people. Earlier in the evening, I briefly joined a group made up of Jewish and Palestinian victims of the conflict. They were clearly calling for an end to the fighting, of course without a very clear plan. Much later in the evening, while I was on my nightly bicycle ride, I came across a large group holding, what looked like, a heated discussion in the middle of Rabin Square (the main Public square in the Tel Aviv). I stopped to try and assess what was going on, and it turned out that it was the same group of both Leftists and Rightists who, in the previous few days of the conflict, had come to blows over demonstrations against the war.

This time, however, they decided that instead of pushing and shoving, they would sit down and talk. The talking went on for hours. I stayed until 3:30 AM. Can I say that they reached any understanding?

No, the distance between the right and Left in Israel is very large, with most of the country in the middle. But, it was positive to see them talking and trying to understand each other instead of resorting to violence.

There was a growing sense of frustration with the Obama administration today among the average Israeli, on a number of different levels. First and foremost, there is the FAA decision to stop flights to Israel. The sense is that maybe it was okay; that yesterday, a bureaucrat on the FAA could decide, based on certain rules, to stop the flights to Israel of US Carriers. But, the decision today to extend it is one that should have been stopped by the White House. The FAA is, after all, an executive agency. Tonight (Israel time), the State Department spokesman spoke about the fact that there are missiles that can reach Ben-Gurion Airport. Why she had to state that is an interesting question.

There is also resentment regarding Secretary of State Kerry deciding to spend all of his efforts on reaching a ceasefire here. In this case, it looks like his efforts are most likely coming in terms of pressure on Israel. Reports, at the moment, seem to indicate that he is pressuring Israel to accept a 5-day temporary ceasefire. Until now, Israel has demanded a permanent ceasefire, fearing that 5 days will only be used by Hamas to improve its tactical situation.

The major questions that Israelis are asking themselves, with so many complicated and difficult problems occurring in the world at the same time (especially an ongoing war in the Ukraine, not to mention the ongoing slaughter in Syria/Iraq where hundreds have been dying every day) is why has Secretary of State Kerry and the Obama Administration have decided that this is the one problem it needs to solve. Why is it is now trying to solve this problem by trying to pressure Israel in different ways to do so? Maybe it is because President Obama seems to have no leverage over the Russians, the Iranians, the Syrians, or the Iraqis and feels that he can exert his limited powers in the one place that he has some leverage- namely, Israel.

Tonight, once again there are rumors in the air that there will indeed be a ceasefire in the coming hours. However, the latest speech by Halid Mashal, the head of Hamas, seems to indicate that while Secretary of State Kerry and others keep spinning ideas regarding bringing about a ceasefire, Hamas does not seem willing to compromise. He stated that Hamas is not stupid and they will not agree to a temporary ceasefire without achieving their goals. Tonight the Israeli government seems to have gotten the message and believes that the time is not ripe for a ceasefire. Kerry in the meantime has headed back to Cairo.

July 24, 2014
DAY 17 @ WAR WITH HAMAS
It Started Well; Ends Less So

The day began positively. My son woke me to tell me that the FAA had changed its mind and, late last night, gave the green light to airlines to resume flights to Israel. Israelis breathed a collective sigh of relief that the FAA was denying Hamas one of their few victories that partially closing our skies. With minutes of the FAA announcement, US Air and United announced a resumption of their services and, by the afternoon, the Europeans had made the same decision. By this evening, the first flights of European airlines started landing and the first of thousands of stranded Israelis began returning home. Also, early this morning, the Israeli TV began showing pictures of a group of 150 Hamas members who were said to have surrendered in mass.

For a few minutes, it was starting to feel like maybe we had turned a corner and this might end at a reasonable point for Israel. Just as I was reflecting on that possibility, I saw on the television screen that rockets had been fired at various points near to us, but not in our exact location. Rather, over a city next door. Since I did not have to go to a sheltered place, I had the unique opportunity to look out of my living room window and watch an intercept taking place in the sky over a neighboring suburb. I was even able to get a picture of the cloud of smoke that resulted from the intercept. I did not have much time to enjoy the picture, however, since within a minute the sirens in Tel Aviv went off. Another round of rockets were aimed at us. Within a minute of going into our secure area, we could hear four booms - two loud ones directly overhead (the intercepts) and two not as loud (the extra missile self destructing). So maybe it was not going to be over so quickly.

As the day proceeded, it became clear that there was a real change. As of 10PM Israeli time, Hamas had only fired 46 rockets at Israel, which is a significant decrease in their average of 120 missiles per day. Israeli army sources report that Hamas has decided to be more judicious in their use of missiles, believing that the current conflict could go on for another two weeks and they want to make sure that they do not run out of missiles. Another two weeks? What am I supposed to tell my son? This is what his whole summer is going to look like, never knowing when the sirens might go off? As I was writing, this terrorist from the Sinai desert (Egyptian territory) fired missiles on the city of Eilat. But, it is going to be Egypt's problem to find them.

Meanwhile, in Gaza itself, the hard work of destroying the tunnels continues. So far, 11 of the 31 tunnels that have been found have been destroyed. The army believes it will take another 4 or 5 days, at minimum, to destroy all of the tunnels that they have found. This assumes that they do not find any more. Israeli troops, other then looking for more tunnels, have been more or less static in Gaza today, with small additional raids taking place to keep Hamas off balance. This afternoon, a tragedy took place at a UN school that was being used as a refugee center. A mortar shell landed in a courtyard and killed15 people. As of this hour, it is still unknown as to who fired the shell.

Meanwhile, Secretary of State Kerry is still in Cairo trying to obtain a ceasefire. Despite the very clear 'no' of Hamas yesterday, Kerry continues to try. The Israeli security cabinet met last night to discuss the situation, but did not vote on any ceasefire proposal. The Israeli government's position at the moment is that it is willing to have a ceasefire as long as it can continue to destroy the tunnels. Israel has also made it clear that it will not discuss a ceasefire proposal until Hamas accepts it. After accepting ceasefire proposals twice, and having Hamas reject them, Israel is not willing to be the first to accept a ceasefire yet again.

Tonight in East Jerusalem and at the Qalandia checkpoint that separates the West Bank and the area controlled by the Palestinian Authority and Jerusalem large scale demonstrations and rioting are taking place. There are reports of deaths and injuries among the rioters who were throwing stones at Israeli soldiers. The demonstrations are at the behest of Hamas who has called on the Arabs of the West Bank and Israel proper to demonstrate in support of Hamas and the people of Gaza – a disturbing development indeed.

July 25, 2014
DAY 18 @ WAR WITH HAMAS
No Ceasefire

Israelis were waiting tonight for word as to whether a ceasefire agreement could be reached. There was both hope and fear -- hope, because Israelis have had enough of a war they never wanted in the first place and fear that Kerry would craft an agreement that would reward a ruthless, terrorist group. Kerry presented a plan that, according to the members of the Israeli cabinet, could have been written by Hamas- and which was, almost certainly, written at least in part by Turkey and Qatar. In spite of this blatant bias in their favor, Hamas was still not satisfied. It should be noted that Israel's security cabinet in its first meeting during the Sabbath in 13 years, voted unanimously not to even to consider the proposal. At this point, Israelis believe that the American Administration is either naive or shockingly two-faced. The video clip of Kerry last week seemed to say it all. Kerry declared that he must do something, that this killing must be stopped. It does not matter who started the war, it does not matter who is responsible, who is right, the most important goal is to stop the fighting.

What has the IDF done to minimize harm to civilians in Gaza?

☑ **Phone Calls**

Thousands of phone calls and text messages were sent to Gaza, warning them of IDF strikes in the area.

☑ **Leaflets**

Thousands of leaflets dropped over Gaza warned civilians to "avoid being present in the vicinity of Hamas operatives."

☑ **Aborting Airstrikes**

The IDF has called off airstrikes when pilots spotted civilians — even when missiles were speeding toward their target.

☑ **Roof Knocking**

These loud but non-lethal bombs warn civilians that they are near a target, giving them time to leave the site.

☑ **Pinpoint Strikes**

The IDF has targeted terrorists with pinpoint strikes, minimizing harm to bystanders as much as possible.

What has Hamas done to minimize harm to civilians in Israel?

☑ Nothing.

Hamas' goal is to kill Israeli civilians.

🛡 ISRAEL DEFENSE FORCES

This is a worthy goal, but to weary Israelis, it is impossibly naive. In Cairo, Kerry's tone remained similar. After announcing that he had not reached an agreement with the parties, he stated, "if only the parties would stop firing they could work out their differences." The average Israeli would probably like to ask Secretary of State Kerry how that approach has been working out with Al Qaeda? Tonight Kerry announced that he would keep working and expressed the hope that both sides would accept a 12-hour humanitarian ceasefire.

Now, there is the additional stressor of Israel possibly being on the verge of a third Intifada (uprising) on the West Bank. The demonstrations last night were the largest and most violent seen in nine years. There was a concern that today, the last Friday of the Ramadan, would be even worse. Particularly worrisome was the fact that many of the army units normally located in the West Bank are now in Gaza. These units have been replaced by reservists who were not given time in advance to prepare.

Thankfully, as Friday has come to a close, the day saw relatively limited violence. Meanwhile in Gaza, the day was also relatively quiet. Israel has now destroyed 15 of the 31 tunnels it had found. It is believed that more tunnels will yet be discovered. Two Israeli soldiers lost their lives. Hamas soldiers seem to be backing away from fighting on a multiple fronts. If the first few days of the ground battles could be characterized by very heavy fighting by Hamas, the last two days has witnessed much lighter resistance. The Israeli government today officially announced that the soldier Hamas had claimed to be kidnapped was, in fact, dead. The number of missiles fired at Israel was much lower. On average, there has been a drop of almost 50% in that number.

On a personal note, the war came a little closer to home last night when I found out that the son of close friends in the United States had been lightly wounded in the Gaza. He was being taken to a hospital in Ashkelon. So off I went at 1:30 in the morning to Ashkelon. I arrived at the hospital at 2:30 AM and found a fully functioning hospital with a hard-working staff situated only a few miles from the Gaza border. The staff was happy that the soldier had a family friend there and they were both extremely helpful and very competent. I am happy to say that the soldiers was only very lightly wounded and after taking some antibiotics and getting a few days' rest will be fine.

One of the somewhat disturbing aspects of my visit to this frontline hospital was the announcement in the elevator as they were transporting the soldier for a test. "This floor includes the department of "mass casualties"." Chilling, to say the least.

its demands would be met. After all, the United States, The U.N. and some of the Europeans believed ending the fighting was the first and foremost goal, and that everything else was secondary.

However, both Hamas and Secretary of State Kerry seemed to have forgotten that there are two additional parties to the conflict – Israel, and to a lesser extent, Egypt – neither of whom agree to Hamas' demands. The Israeli cabinet tonight accepted the UN request for another 24 hours of ceasefire. There is still some hope that Hamas will agree to an extended ceasefire of and then maybe there will be agreement to extend further. That hope seems to have faded some this evening as its clear that the Hamas's military wing is calling all the shots.

The citizens of Tel Aviv awoke this Saturday morning to the first day – in over two weeks – when they did not have to worry about alert sirens going off; or fear missiles falling from the sky; they did not have to dread that their sons, husbands or brothers might be killed or wounded in Gaza at any moment. Today the citizens of Tel Aviv awoke to a temporary ceasefire. Albeit brief, the ceasefire could be felt in the streets of the city. For the first time in weeks the cafes were crowded this Saturday morning. I went down to the beach, (the same beach that last week was almost completely empty), and found it filled with people this Saturday morning. Of course, the beach was not nearly as filled, as it would usually be on a Saturday morning in July. One group was totally absent – the tourists.

Given the quiet, I took the opportunity to chat with people along the beach. I walked away with the sense there was guarded optimism amongst the beach goers. Most decided to return to the shore today because of the ceasefire. Some said they believed the ceasefire would be extended, (beyond initially agreed upon 12 hours). Others believed that the current round of fighting was over. Still others thought the fighting might go on for a few more days until all of the tunnels had been destroyed. None was optimistic that this would be last time we would go to war with Hamas in Gaza.

One of the most pessimistic people I spoke with was a man in his early 70's who has lived his whole life a few blocks away from where we were talking on beach. He talked about the profound hatred expressed by Arabs towards Israel and Jews he has felt throughout his life. Then, he shared a story with me about his father who owned a store in Jaffa, (now part of Tel Aviv, originally a separate town). In 1936, he was warned by a Christian Arab to leave his store and not come back that night. That night, it – and many other Jewish stores – were burned down. His father never returned to Jaffa. He was the most discouraged of all those who I met this morning.

Meanwhile, the Gazans who have now returned to the neighborhoods that were seized by Israel have been shocked by the complete devastation. However, much of that devastation has been caused by Israel's calculated destruction of entire streets under which tunnels ran. Israeli troops report that almost every single house in these neighborhoods was connected to Hamas' intricate tunnel network.

Many people in Gaza have died. The images of the wounded and dead Gazans have entered the Israeli consciousness, but Israelis have no idea how to react. There is no question that it hurts. Still, nobody has any solutions. I heard one opponent of the war ask a former army officer– "Isn't there any way to hit the missiles and not kill anyone?" The answer was an absolute – "No". Four more Israeli soldiers died in Gaza before the ceasefire. The overwhelming majority of Israelis are comfortable with the army using deadly and overwhelming force against targets firing at Israeli soldiers (even if they are hiding among

When I got home, I managed to grab a few hours' sleep and had, as my wake-up call, the sound of a general attack on the Tel Aviv area that did not include Tel Aviv itself but the explosions could be heard. I headed out for some lunch with my son. For the first time, the sirens went off as we were walking. We ducked into a store and waited for the intercept. When it was over, we were able to go into the street and see in the sky the trails of the intercept missiles and, quickly, the point of the impact and destruction of the incoming missile.

Tomorrow is another day, day 19 of a war no one expected and everyone was sure would end quickly.

July 26, 2014
DAY 19 @ WAR WITH HAMAS
12 Hour Ceasefire Ends with Hamas Barrage

I had finished writing this article. It was an optimistic piece, as you will see below. When I first started writing, the 12-hour ceasefire was ending and Israel had agreed to a 4-hour extension. There was hope this extension could be expanded further. However, those hope ended a few minutes ago, when Hamas launched rockets at various parts of Israel. It was hoped Hamas would be pressured by the people in Gaza who got out today and saw some of the devastation – or that Gazans would pressure Hamas to at least keep the ceasefire until Monday, the Feast of Eid al-Fitr. Hamas believes it can achieve more by continuing to fire on Israel. Hamas is only willing to end this war on its own terms, terms that Israel will clearly not accept.

Where does this leave the diplomatic efforts? The answer to that question is unclear. One thing is clear – to date, these efforts have probably hurt more than they helped. Hamas began to believe that civilians). Some are disturbed. Tonight there was a demonstration in Rabin Square, in Tel Aviv. Nearly 5,000 people gathered to call on the government to end the war. This demonstration had to be ended when Hamas broke the ceasefire, thus making it dangerous to have that many people in the public square when the sirens could go off. This was a surreal situation in a very difficult war.

July 27, 2014
DAY 20 @ WAR WITH HAMAS
Ceasefire ... No Ceasefire

Today was one of uncertainty here in Israel. Was there going to be a ceasefire or was there not? Israel stated it was willing to extend the ceasefire last night. Despite Israel's willingness at 8:03, when the 12 hours ceasefire ended, Hamas fired missiles into Israel. At 11:30, a mortar was fired and landed among Israeli troops stationed near the border in Israel and killed one. At midnight, Israel officially offered to extend the ceasefire for another 24 hours. Hamas was silent, but answered in the morning by firing missiles deep into Israel. So it seemed that there would be no ceasefire. Then, suddenly at 1:30, Hamas asked for a humanitarian ceasefire. Israel did not answer immediately, and at 2:03 Hamas fired additional missiles into Israel. Hamas fired missiles into Israel all afternoon.

Meanwhile, Israel told the United Nations that it was unwilling to officially declare a ceasefire, since Hamas had violated every ceasefire thus far. But, Israel added that silence would be greeted with silence. Israeli commanders in the field were ordered, as of 2:00 PM, not to begin any initiatives against Hamas and to only respond to Hamas's actions. The expectations are that, with the Feast of Eid al-Fitr taking place tomorrow, for at least the next 24 hours there will be, more-or-less, a ceasefire. However, as I write this, missiles were fired at several cities in Israel. As of 9:30 PM Tel Aviv has gone two days without missiles. What will happen now is unclear.

At this point, the Israeli public is not in favor of a ceasefire. An unprecedented 87% of the public is opposed to a ceasefire, with 67% of the public believing that the only goal of the project is to end the rule of Hamas in Gaza. I cannot think of any time in Israel's history, since the 1973 Yom Kippur War, that the Israeli public was as united. This is despite what, for Israel, has been the heavy loss of life, over 40 soldiers have died. 19 days of running in to the shelters has convinced the Israeli public that the only solution is to end the rule of Hamas in Gaza. Interestingly, despite the overwhelming support for making the goal of the war to end the rule of Hamas in Gaza, Prime Minster Netanyahu and others in the security cabinet have not adopted that goal. The reason seems clear.

Any attempt to advance deeper into Gaza will, without a doubt, result in an increase in the number of deaths and destruction on the people of Gaza. While Hamas seems to welcome that, the Israeli government has been doing all that it can to not fall into this trap. If Israel has had a hard time dealing with the pictures of death and destruction in Gaza until now, any additional advance will only result in many more worse images. If a diplomatic solution is not found after the current temporary ceasefire, Israel may feel forced to move forward. At the moment, Israel has destroyed 21 of the 32 tunnels and it continues throughout the 'sort-of' ceasefire to carefully destroy the remaining tunnels.

Today, the Israeli press published the full text of Secretary Kerry's peace proposal. Most of the discussion today in Israeli media tried to understand how Kerry could be so naive with his proposals. The overall consensus was that Kerry hurt much more then he helped. First he rushed to come and do something, despite Israel and Egypt begging him to wait. What seems most astounding to Israelis is the fact that Kerry seemed to have developed his plans with Turkey and Qatar, the most extreme of the Muslim states that oppose Israel. Turkey and Qatar are the two countries closest to Hamas and instead of working with Egypt or the Palestinian Authority; Kerry came up with a plan that he developed with them.

One of the leaders of the Palestinian Authority said today that Kerry just does not understand the politics of the Arab world. The Israeli government, who rejected the plan Friday night, was stunned by a plan that talked about all of the specific needs of Hamas, while only referring to Israeli security needs in the most general of ways. In Israel, the belief is that if and when a ceasefire is reached, it will only be reached through the diplomatic efforts of the Egyptians. Tonight there was word that the US has approved an over $20 billion arms deal with Qatar – hmm.

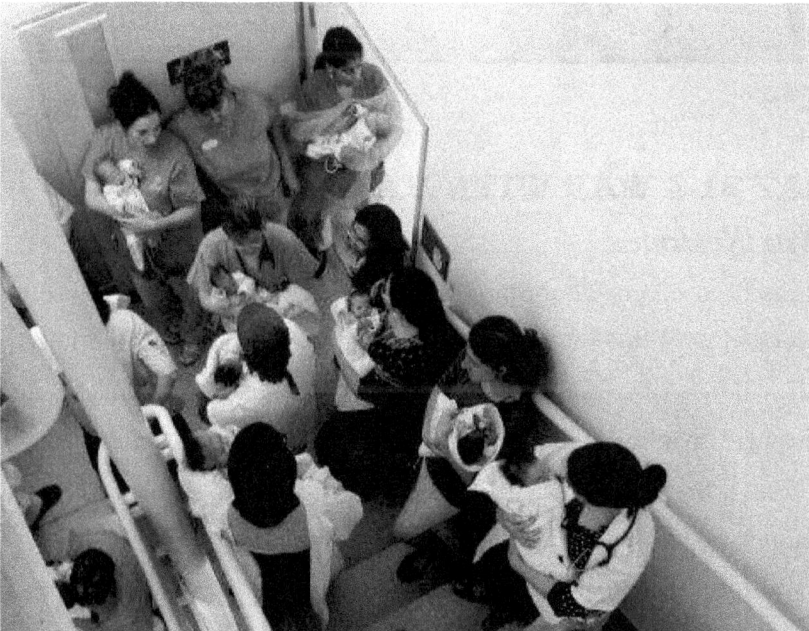

▲ A mixture of hospital staff and mothers holding newborns during one of the missle alerts on July27, 2014

President Obama and Prime Minister Netanyahu spoke tonight. It is interesting almost all the news media have headlined with Obama call for an immediate ceasefire but its one line in a much longer release.

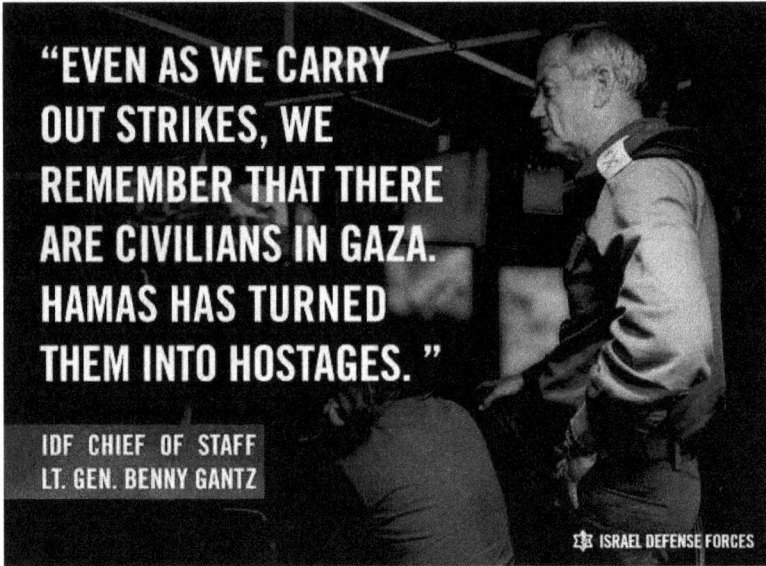

"EVEN AS WE CARRY OUT STRIKES, WE REMEMBER THAT THERE ARE CIVILIANS IN GAZA. HAMAS HAS TURNED THEM INTO HOSTAGES."

IDF CHIEF OF STAFF
LT. GEN. BENNY GANTZ

ISRAEL DEFENSE FORCES

July 28, 2014
DAY 21 @ WAR WITH HAMAS
War Resumes

It has been a very disappointing day here in Israel. In the morning, it looked as if we might be heading into a ceasefire. While neither side had officially accepted the ceasefire, Israel had made it clear since yesterday that it would not engage in any offensive action. In the morning, there was a bit of missile fire, but there was also the sense that it would slowly wind down, at least for the balance of the Feast of Eid al-Fitr. My son asked me what the chances were of this war being over. I told him the chances were 60%. Israel clearly was interested in the end of the conflict and Prime Minister Netanyahu was willing to head in this direction, despite the opposition of many of his cabinet members as well as the overwhelming percentage of the population. (In a snap Internet poll, 87% answered that they did not want a ceasefire until Israel reached its goals.) As the day went on, however, it became clear that, instead of a ceasefire, the situation was getting worse. Hamas increased its attacks. Late in the afternoon, there were reports that Israel had attacked the area of Shifa hospital in Gaza. Israel was immediately accused of the attack, but, as it turns out, the attack was, in fact, a misfire of a Hamas rocket.

One hour later, Hamas fired a series of mortars on a Kibbutz near the Gaza border. The mortars landed in the midst of a group of soldiers, killing four and wounding a larger group. If there was any question as to whether there might be a ceasefire, the Hamas used an undiscovered tunnel to infiltrate Israeli near Kibbutz Nachal Oz. One of the infiltrators was killed, while the others managed to flee back into the tunnel to Gaza. Additional Israelis were casualties of the attack (details have not been released as of this time.) Hamas also fired a rocket tonight, towards the North in the direction of Haifa.

The Israeli government is facing a very difficult dilemma. On the one hand, the people of Israel are demanding that, once and for all, the threat of Hamas be removed. That translates into Israel going much deeper into Gaza. Going much deeper into Gaza means only one thing: much greater civilian casualties in Gaza. This action will result in more photos of dead women and children and additional accusations of Israeli war crimes, even if Israel's actions follow all of the laws of war. The Israeli government has been misjudging Hamas consistently in this war. From day one, Israelis believed that Hamas did not want a war. But, it is now clear that this is exactly what they want. Prime Minister Netanyahu spoke to the Israeli people tonight, but said very little new information. Many were expecting him to announce an expansion of Israel's attacks, and were disappointed when he did not. Netanyahu might, of course, just been misleading Hamas and we may see some surprising actions overnight.

There was a great deal of discussion in Israel today about the telephone call between President Obama and Prime Minister Netanyahu last night. Obama called, among other things, for the "strategic need for an unconditional humanitarian ceasefire." This was seen as pressure on Israel to end the attacks. Tonight, this statement seems absurd to Israelis, especially considering the fact that it requires two parties to agree to a ceasefire, and, as of the moment, Hamas seems uninterested.

There continues to be mutual accusations between Israel and the American government over the attacks on Secretary of State Kerry in the Israeli Press over the weekend. Many American commentators are accusing the Israeli government of directing those attacks. However, as someone who wrote my first article a week ago critical of Kerry's visit, I can say I have spoken to no one in the government. It only requires an understanding of history and diplomacy to understand how unsuccessful and unhelpful Kerry's efforts were.

July 29, 2014
DAY 22 @ WAR WITH HAMAS (Part 1)
A Leftist Call to End Hamas Rule in Gaza

There is talk once again of a ceasefire. I suggest Prime Minister Netanyahu go on National and International Television and say the following: We accept the ceasefire. However, we want everyone to know that this is the last time. If one more missile falls on Israeli's soil, or if one more Israeli is attacked, we will end the war – and we will end this war by ending the Hamas rule in Gaza.

Last night I asked my lifetime friend and mentor, who lives on a *Hashomer Hatzair* Kibbutz in the South, what it means to be a Leftist at time like this. His uncertain answer was the willingness to make sacrifices for peace. So, I guess I am still a Leftist, but a Leftist that believes the time has come to go all the way and overthrow Hamas regime in Gaza. Only if we overthrow this vicious, murderous regime will there ever be a chance for peace. I wrote an op-ed piece on this site, Our Choices in Gaza as the current war was beginning, that we had two choices: 1) Ignore Hamas and rely on Iron Dome to protect us, (my preferred solution), or 2) Go all the way and overthrow Hamas. Though I further stated my fear that half measures would be a disaster for Israel. Unfortunately, we have a Prime Minister who has been given very high marks for his steadiness, for not making rash decisions.

As a historian I cannot help but remember a general in history who reminds me of Prime Minister Netanyahu. That person is Union General McClellan, who headed an army that was far superior to the Confederate army he faced. However, his failure to take decisive action, and his overall unwillingness to take risks resulted in Union defeats – and was ultimately responsible for extending the war much longer than it should have. While our situations are not the same, (e.g. McClellan did not have to worry about world opinion), Prime Minister Netanyahu's failure to make decisive decisions will cost Israel dearly in the years ahead.

Hamas has declared war on Israel. We could have chosen to laugh at them and ignore them, but when we responded, we should have responded like a country going to war– and when you go to war your goal is to defeat your enemy. Yes, there is a chance that if we unseat Hamas something worse could take its place. Having patrolled in Gaza in the early 1980's I do not think that it would be a good idea for us to end up reoccupying the Strip. Still, right now we have a government in Gaza who is continues to plan our destruction, and will take no less. Furthermore, that same government started a war with us. This war was not some sort of mistake. This war was not an event where both sides misjudged the other. This is a case where only one side has made misjudgments – and that is us. Our government and analysts repeatedly said that Hamas does not want a war. Well, they do. And if it's a war they want, we should respond as if this is a war we go out and win. It is tragic that so many civilians are that word collateral damage, and seeing the pictures of the dead in Gaza is terrible. But our Prime Minister, in all of his caution, has wasted at least two weeks of world support amidst his indecision.

It may be too late to take the action needed. I surely hope it is not. I do not write these words lightly. Ringing in my ears is a conversation I had with a soldier who was wounded in this war. When we were discussing the options before us, he asked: "Is it worth 40-50 of us dying to make a point?" The answer to that question is, to make a point – No! Sadly, to change our strategic situation, I am afraid sacrifices will be needed and thinking of them brings me to tears. As someone on the Left, I know there is no choice of making peace, if our enemies know we might withdraw, and allow them to continue to fire. Our enemies must learn that if we withdraw, and they fire, they will lose power. They must know that if we withdraw it is not a sign of weakness – We must do what we have to do from a position of strength otherwise we have no future.

July 29, 2014
DAY 22 @ WAR WITH HAMAS (Part 2)

The day began early in Tel Aviv with the sirens going off at 2:30 AM. Most of my household was still up, though I had gone to sleep at 2:00 AM. It still seems like a dream hearing my son wake me up and soon after finding myself in our secure space. I have a vague memory of hearing the two explosions go off – one louder than the other – and then I went back to sleep. It was a difficult night last night for all of Israel. The IDF announced the death of 6 soldiers, but rumors were circulating that another five might have been killed, bringing yesterday's potential death toll to 10.

By the morning the losses were confirmed. Nine out of the ten soldiers were actually killed inside Israel – five by an attack initiated through a tunnel and four from a mortal shell fired at Israel from inside Gaza. Many Israelis were expecting to wake up to hear the IDF had taken additional actions in Gaza, in response to the actions of Hamas during the so-called "ceasefire", during which Hamas killed 10. They woke up only to the news of the deaths.

Today was another day of uncertainty. Hamas is definitely running low on rockets. There were a few rockets fired at cities in the South today. This evening two rockets were fired at the Jerusalem area – one landed in an open field and the other was shot down. As I write this piece, it is 10:00 PM in Tel Aviv, and the day has been largely quiet.

Well ... seems I wrote too soon. It is now 10:15 PM and there were just attacks on areas South of Tel Aviv. The sirens did not go off in the Tel Aviv, but the intercepts could be heard clearly from my bedroom window. There were additional attacks on Ashdod. Though instead of sending multiple rockets at once, at this point Hamas is sending only one missile at a time.

There are continued talks about a ceasefire, this time (once again) with the help of the Egyptian government. It's unclear whether this effort is much more than wishful thinking. The Head of Hamas's military wing, Mohammed Deif, gave both a victory speech. This could be a speech representing the end of this recent conflict. On the other hand, Deif made it clear he would agree to any interim ceasefire, only a ceasefire that addressed Hamas's grievances. So despite talk of the end of the fighting being near, it is not clear that we are any closer than we were before.

The central story of the day has continued to focus on the Israeli government's relationship with the American government. While the Israeli government has been trying to tone down the conversation, the same cannot be said of the American government. Today, Secretary of State Kerry launched a nearly direct attack on Prime Minister Netanyahu, stating that it was Netanyahu who wanted a ceasefire. The level of the attacks back and forth between the two became so bad tonight that a supposed transcript of the conversation between President Obama and Prime Minister Netanyahu was leaked, and the White House actually issued a statement claiming it was false. A few minutes later, the Office of the Prime Minister issued an identical statement.

Meanwhile, Netanyahu has been under relentless attack – not from the opposition – but from members of his own party, for not taking more decisive actions against Hamas. I do not think there has been a time when a sitting Prime Minister has been attacked during war by members of his coalition to the same extent that Netanyahu has been attacked in the past few days by his coalition members – and even by members of his own party.

Israelis, when they are not looking in the sky for missiles, or worrying about their children serving Gaza, look at the outside world with wonder and concern. Close by, the death continues in Syria. More people keep on dying in Syria (on a daily basis) than in our conflict. Last week alone, 1,700 people died. Why the world does not seem to care is the question that puzzles most Israelis.

The other wonder and concern is directed at Europe. The level of anti-Semitism that has spread throughout Europe in the past few weeks has shocked Israelis. To some extent, I think most Israelis were not surprised by the recent anti-Semitic incidents in France. However, reports of a call in Berlin for Jews to be sent to the gas chambers have deeply shocked Israelis. It is ironic that those demonstrators who are speaking out in favor of a government that claims Israel has no right to exist, are proving to Israelis – in a way that nothing else could – why we need our state.

July 30, 2014
DAY 23 @ WAR WITH HAMAS
The Morality of War

This was a day of treading water in Tel Aviv. It was the 23rd Day of war. On the one had, it has been another day without rockets on Tel Aviv, although rockets were fired at its suburbs. And, as I write this, I could here the sounds of the intercept at any moment. On the other hand, there is a growing frustration that it is not possible to bring this war to an end. To Israelis think it makes no sense that this would go on. Logic says that the missiles have failed to damage Tel Aviv and the short period during which foreign airlines stopped flying to Israel has ended. What does Hamas hope to accomplish? Why does this senseless war continue to go on? Why does Hamas not want a ceasefire? What do they hope to accomplish?

Last night, after the speech by Mohammed Deif, there was hope that this was a victory speech and, somehow, they would agree to a ceasefire. That has certainly not been the case. Hamas and the Palestinian Authority have been fighting over who would represent them at talks in Egypt for two days. The Egyptians seem to be in no rush to bring this war to an end. Hamas also seems unwilling to end the fighting. Hamas has said that it will not consider any ceasefire as long as Israel is in the Gaza Strip, and Israel has made it clear that it will not leave without destroying all of the tunnels- tunnels that keep being discovered and seem to originate ever deeper in Gaza. So the killing goes on.

The gulf between Washington and Jerusalem on the ceasefire process seems to continue. On the one hand, both governments have been trying to downplay the differences, but their surrogates in the media have kept the fire going between the two governments. Secretary Kerry has continued his efforts to bring about the ceasefire, and with that has continued trying to work with Qatar and Turkey, the same players whose involvement is rejected by both Israel and the Egypt. According to a State Department spokeswoman, Israel is aware of the discussions. But, of course, being aware and being supportive are two different matters. Despite those differences there were reports tonight that the US has approved the transfer of US ammunition reserves stored in Israel to the IDF.

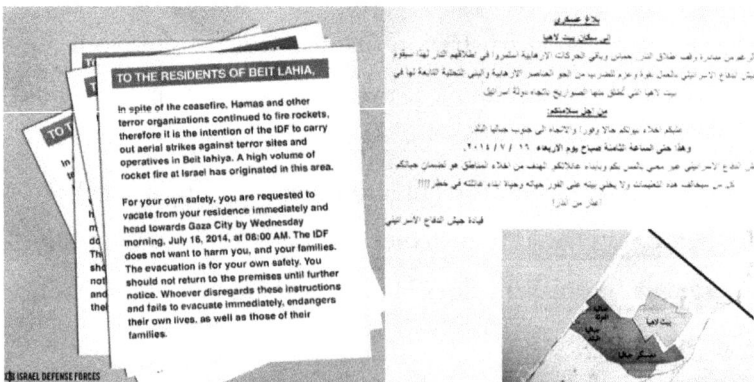

In Israel, there are only limited discussions about the civilian deaths in Gaza. There is a clear understanding that innocent children are dying. But only the right wing who attacks the government- not for the deaths, but for not taking more decisive actions and trying to topple the Hamas government in Gaza. Most Israelis on the Left have been supportive of the war, understanding that it is a war with no easy choices. The settlements surrounding Gaza are all Kibbutzim – kibbutzim who have traditionally been supporters of the Israeli Left. And despite claims in the past years that the days of the Kibbutzim being important contributors to Israeli society are over, in this war when Kibbutz residents are only 2% of the population, 13% of the casualties have been from Kibbutzim. Dr. Noah Efron, of Bar Ilan University, who wrote in a blog post today in the Times of Israel, captured this moral dilemma:

> Hamas is a factory of moral bad luck. Its leaders aim to trap Israel in situations from which only bad can come, either dead Israelis or dead Palestinians or both. They began their barrage of rockets on Israel because they knew Israel would respond, killing innocent Gazans, including kids, along the way. They unleashed their evil because they knew that Israel would, in response, unleash evil of its own.

WHY DID THIS TURN INTO THIS?

BECAUSE HAMAS USES CIVILIAN HOMES FOR MILITARY PURPOSES.

ISRAEL DEFENSE FORCES

Israelis know they are unleashing evil of their own but they prefer that to having their sons killed. Today, Israeli troops are enveloped by an umbrella of supportive fire. Every unit has a fighter bomber flying above to prove close air support, with artillery support ready at a moment's notice. As a result, when ground troops are under attack, the response is overwhelming. This overwhelming response often results in innocent people dying. This is especially the case in a war like the current one, in which Hamas fights from within civilian populations. Israelis do not spend a great deal of time reflecting on the deaths.

When the war is over, most Israelis believe there will be time to sort out the moral ambiguities. Until then, everyone wants to know that his or her sons, grandsons or husbands are going to come back alive and well. They want to know that if their child is in trouble, it will not be a lawyer deciding how to save him, but rather his comrades in arms who will do whatever it takes. Despite the care, three Israeli soldiers died today in Gaza when the house in which they found a new tunnel was blown up while they were inside. A much larger number of Gaza's civilians died, too.

Today, I went to Ben Gurion Airport to pick up one of my daughter's closest friends who was on vacation in the United States when the war broke out. He is a boy/man that I have known since he was 10 years old, and tomorrow, he rejoins his unit in Gaza. Tomorrow, I will have one more person to worry about.

Most Israelis believe that war is immoral. But they ask themselves – what is the alternative?

July 31, 2014
DAY 24 @ WAR WITH HAMAS
When Will It End- Maybe Now?

I just came back from my nightly bicycle ride, and heard the distant sounds of a rocket intercept. I thought to myself– this is kind of late; when I got upstairs there in my inbox was the State Department announcement that a 72 hours ceasefire had been agreed to by all parties. The ceasefire announcement included the sentence "During this time the forces on the ground will remain in place," thus allowing Israel to continue to blow up the tunnels. The statement also referred to the fact that Israel and Hamas will begin negotiations under Egyptian auspices during ceasefire. Of course the ceasefire is suppose to begin tomorrow at 8AM and the sounds I heard in the distance were the interception of Hamas fired missiles at the Tel Aviv suburbs. Israelis are hoping that this ceasefire turns out different then the previous ones.

If this is, indeed the ceasefire that truly ends the war, many dissertations will be written to try to explain why this war went on so long. The proposal is identical to the proposal presented by the Egyptians over two weeks ago, before Israeli ground troops enter Gaza and before so many deaths occurred on both side. Why so many people had to die for no apparent reason is indeed a question worth pondering, and why Hamas accepted this proposal today when they rejected a much better proposal (from their perspective) from Secretary Kerry last week is also mystery worthy of investigation.

The day began very differently. I began the morning by giving a hug to the young man I had picked up at the airport yesterday – on his return from New York; now in uniform he was headed out towards Gaza. Not an easy thing to do. After he departed I read an interesting article in Ha'Aretz Newspaper. It examined all 2,825 rockets that have been fired on Israel in the war and concluded that most missiles were fired on the hour, with the two most popular times for rocket fire being 11:00 AM and 6:00 PM. Tonight at 6PM, the sirens in Tel Aviv went off and the first time in a number of days we received a rude reminder that we are still at war. Not that we really needed this reminder. Not when all of the regular shows on the main TV channels have been superseded by continuous war coverage. And, not when we watch funerals of young soldiers on TV every night.

For a part of the day, there was a sense of optimism in the air. The army reported that no soldiers had been killed or seriously wounded in the last 24 hours. And, although the cabinet authorized an additional draft of 18,000 reservists, most Israelis realized that this was merely a technical move to allow the replacement of some of the reservists who were first drafted. The army announced that 80% of the tunnels had been destroyed and while the South was continuing to be attacked with missiles, they did not seem to be reaching the center of the country. This evening that changed.

First, a missile managed to get through the Iron Dome and injured two people in Kiryat Gat (a city in the Northern Negev), with one seriously so. Then, there were rockets on Tel Aviv, followed by a mortar attack on troops in Israel on the Gaza border. Eight were wounded and, as in all cases like this, the concern is that in a few hours the army might announce that there were fatalities.

Today, there was full cabinet meeting in Tel Aviv. The meeting began with Prime Minister Netanyahu publicly criticizing the members of the cabinet who have been publicly criticizing him. He reminded them that we are at war and thus, this is not the right time for public debates. The public debates have taken place between Netanyahu and members of his coalition who believe that he is not acting strong enough against Hamas. Many Israeli politicians believe that the goal should be to undermine the ruler of Hamas, or at least hurt them beyond the tunnels. Netanyahu has held to his views that Israel's goals must be limited, that Hamas is the enemy that we know, and if the were unseated, Israel might find itself facing the ISIS or some other group instead of

Hamas. The second reason remains the fact that any further moves into Gaza mean more civilian casualties. And, as much as the world thinks we do not care, we do. As I wrote yesterday, when our kids are in danger we will do what we need to do; but, we do not do it with pleasure. We do it reluctantly, and Netanyahu knows this. He knows what the cost of deeper penetration into Gaza would be. He also knows that with 1,400+ Palestinian dead (a substantial number, who are civilians), the world's patience for this war is running out.

So, how will it end? There are now two possibilities. Some in Israel are talking more and more about pulling out unilaterally when we finish destroying the tunnels, something that is expected to happen early next week. At that point, we will pull our forces out of Gaza. And, if Hamas continues to fire missiles, we will leave the response to the air force and the navy. Many people doubt that this scenario can actually work, though, since it is not clear what would cause Hamas to stop the fire. There is talk this evening that Hamas is finally feeling the pressure and may ready for a ceasefire.

Rumors were circulating earlier that Hamas is actually ready to accept a humanitarian ceasefire for 72 hours. That has turned out to be true. After the 72 hours, a long-term agreement can be negotiated under the mediation of the Egyptians. I doubt that too many Israelis believe this will happen. Even the news anchors interrupt the correspondents when they report the possibility, commenting, effectively, "you're kidding." Maybe this time it is real. But we have been here before. We are in the 25th day of the war, in the second week of the ground war. At some point, this clearly will end. I must admit, though, that I never believed this war would go on for so long.

August 1, 2014
DAY 25 @ WAR WITH HAMAS
Israeli Soldier Captured

I am tired of writing these columns, as I know I am beginning to sound like an endless audio track. The day begins on a positive note and then things go wrong, and that is what happened yet again today. When I awakened this morning, I was thinking that today's diary entry would be a summary, as a ceasefire finally seemed to signal the end of the fighting. By the time I returned from a brief foray to the local supermarket, it was clear that things had already gone terribly wrong.

First, my son informed me there had been an alert near Gaza, I told him not to worry as it usually takes a few hours for a ceasefire to be implemented fully. Then reports came in of heavy firing in the Rafah area. My first inclination was to believe that there was some minor local incident. Then, the television broadcast began showing the aerial attacks and it was clear that something had gone very wrong. The intensity of attacks was unprecedented. My heart sank as I posted to Facebook that something terrible was transpiring.

Soon, the full story emerged. An hour-and-a-half into the ceasefire, a Hamas suicide bomber emerged from a tunnel next to a group of Israeli soldiers on the outskirts of Rafah (on the Gaza-Egypt border) and blew himself up, killing two Israeli soldiers. Other fighters emerged from the tunnel and snatched a third soldier, Lieutenant Hadar Goldin. It's not clear whether Goldin was alive or dead or dying when he was captured. At that moment, the IDF deployed its standard tactic -- code-named Hannibal -- designed to do whatever is necessary to stop kidnappers from escaping with a live prisoner. Massive firepower was used in order to stop the terrorists from getting away. Unfortunately,

Hamas has constructed a staggeringly large number of tunnels under that area and the attackers apparently escaped.

For most of the day, I was consumed by a combination of sadness, anger and frustration. At moments, I found myself with tears in my eyes This was not going to end today, it is not going to end tomorrow, Hamas really does not want this to end, and we certainly cannot allow it to end this way. Israel had planning a unilateral withdrawal from certain areas of Gaza in another two days, once the tunnels were all neutralized. Instead, the soldiers whom I thought were no longer going to be in danger are again in danger. More people in Gaza are going to die, as we either try to recover the soldier or go all the way to do what it takes to topple the Hamas government.

The Israeli government has a difficult decision to make tonight. They have to make a choice whether to expand the current action and modify its goals to include the destruction of the Hamas government or to continue the more limited goal of destroying the tunnels. After today's attack, I do not see how the Israeli government will be able to agree to any new ceasefire for the foreseeable future.

Why Hamas decided on this course of action now is yet to be determined. There are a number of potential explanations, the first being that the military wing of Hamas is no longer listening to the political wing meaning that although the political wing accepted the ceasefire, the military wing did not. Another possibility is the military wing decided to try just one more attempt to get the advantage they had not achieved until now: the successful capture of an Israeli soldier. They may have decided that this goal was too important to allow the death of a few hundred more Palestinians get in the way.

Tonight, the Obama administration made some of the strongest statements yet in support of Israel's position. Obama stated in a press conference that if Hamas wants this situation resolved, *"That soldier needs to be unconditionally released"*. He went on to say, "Israel has a right to defend itself. No country would tolerate missiles every 20 minutes, or tunnels used for terrorist attacks." Obama went on to say that after today's kidnapping, it is going to be very hard to reach a ceasefire, since no one is going to be able to trust Hamas.

These sentiments align quite closely with what almost every Israeli is feeling tonight. Gaza has been a problem for Israel since the 1950s. But it had seemed to be a containable problem. Tonight, after the kidnapping and the years of missile fire, there is a general feeling some way must be found to permanently contain the cancer that is Hamas.

▲ A pro-Israeli sign supporting Operation Protective Edge (in Rishon L'Tzion)The sign reads *"Strong in the hinterland, Winning on the front"*

August 2, 2014
DAY 26 @ WAR WITH HAMAS
Netanyahu Announces War to Continue

Hamas finally found a way to get to the average resident of Tel Aviv. A few days ago, they attacked at 2:30 in the morning. Much of Tel Aviv was still awake, so it did not bother people too much. But, this morning, they fired on Tel Aviv at 6:00 AM on a Saturday, waking up much of Tel Aviv hours before they normally awake on a Saturday. I took a walk at 7:00 AM this Saturday morning, a time at which Tel Aviv is usually deserted, and the coffee houses were crowded with people whose sleep had been disturbed.

Prime Minister Netanyahu spoke tonight to the Israeli people. Most commentators were expecting him to announce the end of the mission. Instead, he stated that Israel would continue until it brings about quiet to the South and the country at large. He stated that almost all of the tunnels have been destroyed and the last ones will soon be destroyed. Thus, Israel will begin redeploying its troops and will not be withdrawing them. Israel will deploy its troops to positions that meet the country's security needs. Netanyahu's remarks, and those of Defense Minister Ya'alon, reflect the decision made last night by the security cabinet, which met for five hours to discuss the next steps. It was decided not to change the goals of the "operation" as the government is still calling it (instead of war). On the other hand, it was decided not to enter any negotiations with Hamas regarding a ceasefire. Israel is no longer interested in reaching any ceasefire agreements with Hamas, since Hamas has violated every agreement to date. Netanyahu is clearly saying to Hamas that they are not going to get anything politically, since we can live with War of Attrition.

The Israeli army has officially declared Lieutenant Goldin dead. Forensic evidence found at the scene proved that he was dead before his body was taken. A committee of Rabbis agreed with the army findings and officially declared him dead, and the family was notified tonight. When the attack took place the deputy commander of the unit ignored standing orders and entered the tunnel to try to stop the Hamas fighters. But, they were not able to capture them and realized that his men in the unknown tunnel were essentially sitting ducks, and pulled back. Hamas claims to have no knowledge of the status of Goldin. The Israeli army believes Hamas is lying.

Israelis are very divided on what the next steps should be. 70% of Israelis believe that Netanyahu is doing a good job. 32% of Israelis believe that Israel should go all the way and topple Hamas. 60% of Israelis believe that if the war ends with the destruction of the tunnels, then we will have won.

The problem facing Israel is that its two choices are very difficult. Toppling Hamas is a major effort that will cost many lives. Israelis are very divided about whether that price is worth paying. There are those who say that there is no way to topple Hamas, and only a political solution can solve the problem. But, there are others who believe that no political solution can be found as long as Hamas is in power at all. Beyond that, there remains the fact that the cost on both sides will be very high. In the last week, the support in Israel to go all the way has gone down. The general sense is that this will end and both sides will prepare for the next round. More then sixty years of unwanted wars have made Israelis fatalistic.

August 3, 2014
DAY 27 @ WAR WITH HAMAS
Israel Moves Most Troops Out of Gaza

Today, Israel began implementing the policy it announced last night. Israeli troops began to withdraw from Gaza and redeploy. The Israeli government decided not to expand the operation after destroying all of the tunnels that the army was aware of. The problem with this policy became very clear at 5:00 PM today, when the sirens went off again in Tel Aviv. I was on the phone with my daughter and just said "missiles!" and hung up. This time, the wait seemed forever. But finally, the two booms came that signaled a successful intercept. The second one was almost above us. It was a day of very heavy rocket fire by Hamas, especially in the South. In total, Hamas fired as many rockets today as they had fired earlier in the war. Lately, their long-range rocket fire has decreased significantly and by all accounts; they are running out of long-range missiles. However, they still have 1,000's of short-range missiles that can hit the cities in the South. Today, they were trying to show that they could not be ignored.

Israelis are very confused tonight. According to one poll, 6% of Israelis say that they oppose the ending of the operation. In that same poll, 59% of the people are very happy with the actions of Prime Minister Netanyahu who has ordered most of Israeli troops out of Gaza. A different poll shows that 50% of the public does not want the operation to continue.

The average Israeli in the street mirrors the same confusion that the polls show. On the one hand, they are appreciative of the fact that Prime Minister Netanyahu has been cool under fire. He has not acted hastily or taken too many risks. On the other hand, most Israelis do not understand how the current policy will bring about an end to the missiles. And, if it does bring about an end tomorrow, what will stop us from reliving this month's events next year. Israel's bitter history with Gaza goes back to the 1950's and nothing seems to change that. The Israeli government is effectively announcing that silence will be met with silence - aka a ceasefire, which is what Israel offered Hamas before this all began. The only reason to believe that Hamas may end this now is that they are running out of rockets, their people are homeless, and despite the fact that Israel is allowing in food and medicine to Gaza, the people are running out of food, water, and other essentials. Only if they end the firing will there be a way to solve these problems. Will they? It's not clear. As I have written before, Israel has consistently misjudged Hamas.

The hope in Israel is that negotiations will eventually begin in Cairo; not right away, but in a few days, at which point Hamas will be desperate for an agreement to help the people of Gaza. They will then be willing to go along with an agreement that severely limits their ability to rearm. Wishful thinking? Probably.

Today was another sad day of funerals. Four funerals were held in Israel for soldiers who fell. The funeral that the Israeli public felt most strongly about was that of Lieutenant Goldin, the soldier who was presumed kidnaped on Friday. It turned out that the Israeli army was able to retrieve enough of his body to both confirm his death and hold a funeral under Jewish Law. On Saturday night, just hours before his death was announced, his family appeared on TV begging that the government not leave him behind in Gaza. Thus, when his death was announced last night after midnight and the funeral was held today, the Israeli public at large, together with the 15,000 who attended the funeral, all shared the family's pain and loss.

Israeli parents will be sleeping just a little bit better tonight, as the broad majority of troops have been taken out of Gaza. The chances of troops being killed or wounded tonight will be much lower than previous nights past. The number of innocent civilians killed in Gaza will also decrease, and that is just as important. Israelis, however, know that 20% of Israeli soldiers killed in this operation died on the Israeli side of the border due to cross border attacks. Palestinians know that they were dying before Israeli troops entered by ground, from air strikes. So while less people will die tonight in the Gaza area, people will continue to die on both sides for no reason, until the missiles stop. And when they do, the airstrikes will end, too.

August 4, 2014
DAY 28 @ WAR WITH HAMAS
Ceasefire – Maybe Finally?

A 72-hour ceasefire has been accepted by both sides – starting tomorrow at 8 AM. Israel announced a unilateral ceasefire during the day, but Hamas did not accept it and continued to fire. By tomorrow morning, the last of the Israeli troops will be out of all the areas of Gaza (with the exception of points just along the border). The Egyptians had warned Secretary of State Kerry last week that there was no chance of a ceasefire holding as long as Israeli troops were in Gaza. Sure enough, they were correct. Now the time is ripe for a ceasefire. Of course we have been here before. So until the guns and missiles are silent for a significant period we will not know if the ceasefire is real.

Today was the 28th day of Israel's War with Hamas. If you would have asked me or any other Israeli four weeks ago if this could go on for almost a month, we would have said there is no chance, but I will get to that a little later.

Israel finished blowing up the last of the tunnels today – the tunnel that was discovered when Lieutenant Goldin was killed and his body taken. It turns out that this tunnel went 1.5 miles into Israel and another mile into Rafach towards the coast. New details of that event came out today when it became clear that the three soldiers were not killed by a suicide bomber, but rather by an attacker who advanced until being killed.

The people of Gaza desperately need a ceasefire to go about rebuilding. Of course, how much rebuilding will take place is dependent on the political agreements that can be reached. However, the arrangements that can be worked out are unknown, and only time will tell.

Today's news from Gaza was interrupted by two lone-wolf terror attacks in Jerusalem. In one attack, a 19-year old Palestinian resident of East Jerusalem drove a crane into a bus, toppling it onto a pedestrian who died. A passing policeman then shot the terrorist. An hour later, an unknown gunman on a motorcycle shot a soldier at a bus stop. These are the sorts of events people have feared. Hopefully, if the war comes to an end, these attacks will not spread further.

A new song written by General Yoel Galant (res.), who had been the Southern Commander, was played on the news this morning in memory of the soldiers who died in this war. The song played as the faces of the soldiers flashed across the screen. While I first found only mild tears in my eyes, these tears continued and turned into a flowing stream. Why did all of these young men have to die? What has been accomplished? As the war was beginning, I wrote an article, entitled: "Our Choices in Gaza", stating that we had only two good options – *Option #1:* Ignore the rockets and show the Hamas how weak they were and how our technology (Iron Dome) made all their efforts useless. Or, *Option # 2:* Do what other nations do when attacked, namely declare war and end the Hamas rule – whatever the cost.

Unfortunately, we executed neither of these options. All the young men whose pictures I saw on the screen died. All the Palestinian civilians, whose only sins were being born in Gaza, are also dead. Why? Of course, the answers do not come easy. I have been living in this country on and off (more years off) for almost 40 years. I attended my first rally on behalf of Israel in Washington when I was in the 7th grade during the Six Day War in 1967. I visited Israel during the War of Attrition, attended the reburial ceremonies of some of the Israeli soldiers during the summer after the Yom Kippur War in '73 and did my army service here. I spent time patrolling Gaza, and watched two of my children do their army service already. After all this time things are not getting any better. Why do we do this? There must be a better solution.

I spent some time in Poland this past Spring and recently completed an iPhone/Android app guides to major sites in Poland (with special attention to the Jewish History of Poland). I visited all of the Death Camps and was, of course, reminded why we are here. What happens when Jews do not have an army to defend them? Zionism, as envisioned by Theodore Herzl, should have solved the Jewish problem and provided a safe place for the Jewish people. However, unfortunately, the ancestral Jewish home Palestine/Israel was not an empty place. And, until the other nations who live here (the Palestinians) are willing to truly compromise, I fear the cycle of violence will continue.

On a final note… As I finished this article, the buzzer to our apartment rang. The soldier I had picked up at the airport last week came home. He was given a 48-hour leave. I guess the war may really be over.

August 5, 2014
DAY 29 @ WAR WITH HAMAS
The Day This War Ended
As of 8 AM this morning this summer's war between Israel and Hamas hopefully came to an end. A 72-hour ceasefire went into effect. In the minutes before the ceasefire, Hamas fired a barrage of missiles into Israel. One landed on a Palestinian house outside Bethlehem.

Negotiations for a longer ceasefire and temporary political agreement begin this evening in Cairo. No one is expecting the war to resume at this point. As for the political agreements, it is too soon to know what to expect. Israel will demand some sort of demilitarization of the Gaza Strip in return for rebuilding Gaza and allowing more goods in. Hamas, on the other hand, is going to demand free movement of goods and people in and out. It is not clear what the agreement to be reached will look like. Regardless of the agreement, both parties will come out of this war as losers have lost too many men, women and children.

However, when historians look at the war some time in the future, the deaths will not be considered important. What will be important is the relative political positions of the parties after the war, and how long any agreement that is reached will manage to remain enforced. The key player this time is Egypt, whose strategic interests are more aligned with Israel than with Hamas. The current Egyptian government sees Hamas as an extension of their enemy the Muslim Brotherhood. Egypt controls the major border crossing with Gaza at Rafah and they are the ones who have kept it largely closed. Only in the coming days and weeks will we know the political outcome of this war.

In terms of numbers ... the war, based on IDF figures, can be summarized as follows. In Gaza, 1,798 people were killed of which, according to Israel, 800-900 were combatants. The remaining balance of fatalities were civilians. In Israel, 64 IDF soldiers and 3 civilians were killed. Hamas fired 3361 missiles of which 584 were intercepted, 115 landed in populated areas, 2542 landed in open areas and 120 fell on to Gaza itself. All of the rockets fired at the Tel Aviv area were intercepted.

The only clear winner of the war was the technology of the Iron Dome System. Because of this technology, and other defensive actions taken by Israelis, what could have been a very devastating war, in terms of casualties, was just the opposite. The Israeli economy, on the other hand, suffered much more. Across the board, sales in the country were down by about 20%, with business in the South, of course, losing much more money then the rest of the country. The government will be reimbursing the businesses in the South up to a line of 40 kilometers from Gaza. As for the rest of the country, only businesses in the tourism fields will receive compensation.

Tonight, Israelis, by and large, understand that this war in Gaza has clearly hurt Israel's public image in the world. The governments of the world seem to better understand Israel's need to act than the people around the world. The media's pictures of dead civilians horrified much of the world (as they should), however the reasons explaining those photos were lost in the text that most people did not even bother to read. Israelis themselves have been removed from some of the moral dilemmas of the war. Israeli television does not show very much footage of the death and destruction in Gaza and instead focuses on the issues that the Israeli troops face. In some ways, Israelis have been living in a parallel universe to the rest of the world, and thus do not understand what has upset so many people.

While there is no question that the events of this last month have traumatized the children of Gaza to a degree that is unimaginable, the children of Israel have no doubt been affected deeply by the sudden sounding of the sirens and, in turn, running for shelter. A decade ago, Israel produced a generation of kids traumatized by bus bombings. Now Israel has a new generation with its own trauma.

Tonight, Tel Aviv tries to return to normal. At the moment, it seems like a feeble try. But in the upcoming days and weeks the war will pass in to history. The people of Tel Aviv will instead return to worrying about the cost of food, religious coercion and other areas of concern. And, everyone will continue to try to be a part of the next great start-up. Hopefully, this will be my last update on this war and I can return to writing about other aspects of Israel.

August 7, 2014
DAY 31 @ WAR WITH HAMAS
War May Resume in the Morning

There are less than 9 hours Left to the initial Hamas-Israeli ceasefire. Israelis were quick to celebrate the seeming end of the war by returning to the shopping malls. Last night the cafes and bars in Tel Aviv were full – like they have not been since before the war. Many members of the kibbutzim surrounding Gaza have returned home. The call centers of the major travel centers and airlines were having a hard time keeping up all of the inquiries from Israelis trying to book trips in what is Left of the summer vacation. The Civil Command informed everyone they could return to their regular routines. Reservists have been released from service and enlisted army troops have been reassigned. I even wrote that I had posted what I hoped would be the last installment of my diary during this war. The only problem is we all forgot to ask Hamas what their plans would be.

This afternoon Hamas declared that if its demands are not met by tonight it would resume firing missiles at Israel tomorrow morning at 8:00 AM, when the ceasefire ends. According to reports over the course of the day, Hamas has stiffened its position and is completely unwilling to compromise – not on the essential issues, nor in terms of extending the ceasefire. The situation is further complicated by Egypt's position. The Egyptians are in no rush to make concessions on the question of the Rafah crossing. Their position remains that Rafah is a bilateral matter – between Egypt and Gaza – and has nothing to do with an Israeli-Hamas ceasefire.

Throughout this conflict Israel has misjudged Hamas. Before the war began, all of Israel's greatest analysts said that Hamas had no interest in going to war with Israel again. They contended that Hamas had everything to lose and would not want to risk a war. Israeli analysts, and with them the rest of the world, still seem to be making the same mistake. There are 1,800 dead; 32 tunnels destroyed, and only 30% of Hamas missile supplies remain. Logic would suggest Hamas should accept a ceasefire and get whatever it can get. However, there never was any logic in the current war. From the start, Hamas felt it was losing on all fronts, and took the chance that a war would reshuffle the deck and leave them in a better position.

Unfortunately for Hamas, they were counting on three main strategies: their rockets, their tunnels, and world opinion. Thanks to Iron Dome, their rockets totally failed to do any more than limited psychological damage to Israel. Their tunnels were used successfully for small scale attacks against Israeli troops, however, failed in their main goal to kidnap Israeli soldiers. Lastly, while the destruction of buildings and deaths of civilians in Gaza has certainly effected public opinion and turned additional people against Israel, with the massive bloodshed that has plagued the Arab world in the last two years (especially in Syria), it is very hard for the current deaths to effect world opinion in the same ways it would have in the past.

What does Hamas want? Hamas has demanded that the distance Gazan fisherman are permitted to fish be increased – a demand to which Israel has agreed. It has demanded that the salaries of its employees be paid – again, Israel agreed. It has demanded allowing the free flow of people through the Erez crossing– on this matter Israel has only offered to be more flexible. Hamas has demanded the free flow of cement– To this demand Israel has agreed only under the condition of international control. Hamas has further demanded the free flow of "dual use" items – Israel has said no to this demand. Israel has also responded with an absolute no to Hamas's demands for a port and airfield. Hamas also is insisting that its members who were arrested in the West Bank be released – Israel has also responded no to that demand.

Israel has offered to increase the flow of humanitarian aid, help with the wounded, and with anything else it can in the area of humanitarian aid. It will be very hard for Hamas to explain to its people why it went to war and did not gain anything. Therefore, Hamas is threatening to resume the rocket fire at Israel, even if it makes no sense and they can gain little. Tonight its spokesman warned of a long war- a war of attrition with Israel, if its demands are not met.

At this point, I would predict the chances of war resuming tomorrow morning are at least 50/50. Someone might still pull a rabbit from their hat overnight, but I would not bet on it. I hope that if Hamas does resume its fire, Israel will do its best to ignore it. While that will be hard to do, unless Israel is willing to remove the Hamas rule from Gaza, there is little Israel can do, but stand aside and let Iron Dome do its job. We should order a few more batteries and hope for the best. If we ignore them, Hamas might finally realize there is little they can do, but accept the current terms on the table.

August 8, 2014
DAY 32 @ WAR WITH HAMAS
War Resumes

I stirred a few moments before 8 AM wondering what the day was going to bring. Would Hamas resume firing as they threatened? A quick check of my iPad showed that mortars had been fired about four hours before. And yes, a very few minutes later at 8 on the dot, Red Alerts started appearing. Rockets were being fired from Gaza at Ashkelon, as well as at towns and Kibbutzim around Gaza. The missiles continued to be fired all morning, wounding three people. At 10 AM, the Israeli government gave the word to start returning fire and the air force subsequently began attacking a number of targets in Gaza. In reality, however, there were really no new targets to hit. Tel Aviv streets this morning were empty. I went in to my favorite coffee shop and at first, was the only customer. The owner told me that business was down 50% this month. As of 9PM tonight, Hamas has not fired on Tel Aviv or other parts of central Israel. The major question that Israel faces tonight is whether Hamas was firing just to try to improve their negotiating position, or are they really committing to enter into a long-term War of Attrition? No Israeli commentator seems willing to hazard a guess as to the intentions of Hamas. Clearly, Hamas was surprised that its demands were not met and it still seems to be having a hard time fully accepting the fact that the Egyptian government is not sympathetic in the least to Hamas' demands. It's not clear what will come next. Hamas seems unwilling to compromise and Israel is not likely to give in to Hamas demands anyway. For the first time, criticism in Israel is being voiced against the government. Throughout the crisis, Prime Minister Benjamin Netanyahu was given high points for his management of the war. Until now, only the right-wing had criticized the government for not being aggressive enough and not mounting an all-out effort to overthrow Hamas. Now, criticism is more widespread. Why was the country told it was over? Those who returned in the last two days to kibbutzim situated near Gaza are furious that they came back with their families and now the rockets are falling again. Cities around the country had to order their air raid shelters reopened after closing them once the ceasefire went into effect. The realization is dawning that the government is continuing to misread Hamas.

There is clear disappointment in the apparent failure on the part of the government to have any attainable goals. All of this has been brought into sharper focus when in the last two days, the highest representatives of the government spoke and stated that 'Hamas would not dare restart the war, and if they did, we will hit them hard.' Instead, Israelis woke this morning to missile fire on the Southern part of the country, and watched as the military's only reaction was to initiate the same response as before: retaliatory air strikes. Far from the overwhelming force so casually promised by the government. It is now clear that the government has failed to understand that goals of Hamas from the beginning. Revealed tonight was that intelligence had been obtained to suggest that Hamas was planning a July War to fight what they call the blockade of Gaza. In a chilling replay of events before the 1973 Yom Kippur War -- when the intelligence that Egypt planned to attack existed but was ignored because it did not fit what strategists thought was going on -- it appears that Israel ignored the intelligence because the decision-makers believed Hamas was weak and uninterested in fighting. In truth, however, Hamas carefully planned this war. While it may not have achieved Hamas' goals, the evidence suggests that Israel's government never developed its own set of strategic goals and that its actions reflect that lack.

August 9, 2014
DAY 33 @ WAR WITH HAMAS

Today is the 32nd Day of Israel's War with Hamas – or is it? In Tel Aviv there is no real sense of war. The beaches were crowded, though they did not appear to be filled with the usual number of tourists for an August day. However, Israelis did crowd in to replace them. My son turned to me, as we looked at the hundreds of people in the water standing at the wave line and said: "boy they must all be pretty sure there isn't going to be missile attack today." This was the case – at least around Tel Aviv. No one's son, boyfriend or husband was in Gaza, and it was a fine summer day. Of course, a mere one hour drive to the South (in the areas immediately around Gaza) no such calm prevailed.

Over the course of today 30 missiles and mortars landed in that area. Israeli news broadcasts repeatedly played a clip from last week showing the Chief of Staff stating that the quiet has been reestablished in the South: "and the farmers can now go back to tending their fields and the flowers will grow." A statement that I am sure the Chief of Staff wishes he could erase. Israel has responded to the renewed missile fire with selected attacks on missile targets. Five more people have died in Gaza as a result, including two who were targeted by an Israeli drone and killed (after being caught driving away from a missile site on motorcycle.) This response has far from satisfied the residents of the South, who feel that their interests are being ignored. Once again, they are being fired upon and the rest of the country goes about its routine.

There have been contradictory reports coming from Cairo all day. Anything I write now could change at any time. As of now, Saturday night (Israel time), the Israeli delegation is not returning to the table for talks unless Hamas stops firing its missiles. The Israeli position is that there will be no negotiations under fire. At the moment, Hamas's position is the opposite – so much so that they are threatening to withdraw from the talks if Israel does not return to negotiate. Hamas has also threatened to resume its fire on Tel Aviv tomorrow, if their demands are not met.

Most observers are coming to the conclusion that we are in for a war of attrition, of an undetermined length of time. I think Hamas may be misjudging the situation – regarding who can better withstand a war of attrition. They came to (what from there standpoint is) a reasonable conclusion that Israel was interested in a ceasefire, and that their rule in Gaza was not threatened by Israel. Since the Israeli government made it clear it did not want to go all the way. I believe Hamas concluded it was in their interest to extend the fire, in order to improve their position. However, that assumption led them to the wrong ultimate conclusion. Israel is in a much better position than Hamas to withstand a war of Attrition – and certainly more so than the people of Gaza. The tourist season is over. At this point, the economic losses in Israel will be limited. People will go on with their lives. Whatever psychological damage has been imposed on a generation of our kids has already made its impact. Currently, it is the people of Gaza who are not getting the aid they require. It is the people of Gaza who will suffer if this war goes on. It is via Israel that Gaza gets all of its food and fuel. Every day – even during this war – trucks head from Israel into Gaza bringing needed food and medicine.

One final thought – on the failure of secular leaders to fully understand a religious based military/political movement. Last night I wrote about the failure of the Israeli government to understand the intentions of Hamas. At this point, everyone agree that this was and remains a very large problem. Though this is clearly not a problem that is limited to Israel. This morning, in Washington, President Obama issued a statement and answered questions regarding Iraq and America's new air campaign. In his answer to one of the questions President Obama stated:

Did we underestimate ISIS? I think that there is no doubt that their advance, their movement over the last several months has been more rapid than the intelligence estimates and I think the expectations of policymakers both in and outside of Iraq.

I would suggest this continual misreading of these organizations is a systemic problem that, in recent history, goes back at least to the Iranian revolution. Intelligence analysts have been repeatedly caught flat-footed in trying to understand religious movements. Policy makers should at least be cognizant of these failures when they make their decisions.

August 10, 2014
DAY 34 @ WAR WITH HAMAS
Missile Fired at Tel Aviv Minutes Before Ceasefire

Today is day 34 of Israel's war with Hamas. It was a quiet day until hours before a new ceasefire was to go into effect. For most of the day was another day of semi-war. Hamas "only" targeted locations close to Israel's Gaza border and the IDF limited its response to a small number of attacks. But as the clock neared midnight Hamas began firing at other cities. At 10:00 PM they attacked Ashdod, at 11 Ashdod again and then Ashkelon, finally as the clock neared midnight at 11:58 and explosion could clearly be heard in Tel Aviv. No sirens went off, most likely a missile landed off course in open areas or at sea. As of now, a few minutes after midnight, a 72-hour ceasefire is supposed to be in effect. Hamas has, (at least for the moment), climbed down from its tree and agreed to the ceasefire. Israel will return to the negotiating table tomorrow in Cairo – if indeed the firing completely ends – as has been promised by Hamas.

If the war does actually end with this impending ceasefire (for which there are no guarantees and many believe is unlikely), Israelis will be left very confused. There are many questions left to answer. Did we win or not? Is Hamas weaker or stronger? Did our government do a good job or a bad one? These are the standard set of questions asked at the end of every military campaign. The answers often break down along traditional political views.

However, this time is different. This time Israelis are confused about matters beyond the usual events that lead toward the end of a war. The landscape in the Middle East has been changing so rapidly over the course of the last two years – enough to legitimately give any average spectator a case of whip lash.

The usual calls to turn the terrible events of the past month into an opportunity to achieve a breakthrough are already being sounded, "let's turn lemons into lemonade", as the saying goes. Though average Israelis, the ones who live in places like Tel Aviv, the ones who oppose – and always opposed – the settlements, look around the Middle East today and ask themselves if now really is the time to take risks.

Tonight, on one of the most distinguished Israeli news shows, a professor attempted to explain what is going on in Lebanon today between ISIS and the rest of the Sunnis, the Shiites and Christians. I have an MA in International Relations with a specialty in Middle East Studies from a prestigious University, but was still unable to understand what is truly going on in Lebanon. Average Israelis see a Middle East in which the United States is finally, and very reluctantly, intervening on behalf of the Kurds (a strong ally of the U.S. and a nascent democracy) before they are beaten by the most extreme Muslim force in history ... a force that makes Al

Qaeda seem moderate, an entity which makes the extremist Iranian regime appear moderate and sane. Then, they ask themselves: how can this be the time to make concessions? How can now be the time to take risks?

The war in Gaza has been a wake up call for many liberal, Left-leaning Israelis in other ways as well. Most Israelis, who supported Israel's withdrawal from Gaza, did so believing that once we withdrew, if they fired on us, we would be justified in using as much force as was necessary to stop them. Over the years that notion has proven to be patently untrue. Despite over a decade of constant missile fire over our border, the world has repeatedly accused Israel of using disproportional force. Moreover, most Israelis were deeply shocked at the level of hatred displayed towards Israel and Jews during numerous anti-Israel held during this war; a war Israel clearly did not start, nor did it want.

As a result, Israelis end this war (if the end is in sight) perplexed. What do people expect us to do when our cities, and even our capital, is fired on repeatedly with missiles? Is the hatred of our critics so strong that the only thing that will satisfy them is our death?

One final note ... Israeli media gave extensive coverage today to Thomas Friedman's interview with President Obama. Those "in the know" were surprised to see President Obama acting as a political commentator. The President told Friedman that Prime Minister Netanyahu is politically too strong to make concessions. Israelis were shocked to see the President of the United States so misread his Israeli colleague. Historically, only Israeli prime ministers who were politically strong (e.g. Menachem Begin, Yitzhak Rabin and Ariel Sharon) have been able to make concessions. Furthermore, though Prime Minister Netanyahu's poll numbers were high this past week, his political situation within his own coalition is anything but strong. Even if Netanyahu was inclined to take significant risks for peace, he would find it very difficult to do so given his current fragile political coalition.

August 12, 2014
DAY 36 @ WAR WITH HAMAS
Day Two of the Three-Day Ceasefire

Today is the second day of the agreed upon three-day ceasefire. Hamas has threatened not to renew the ceasefire when it expires – unless there is an agreement in place. I just heard Finance Minister Lapid warn Hamas, once again, that they had better not resume firing – because this time (as opposed to last time) we will respond with great force ... Of course, that is the same thing Lapid threatened before. Every Israeli commentator whom I have heard questioned today refused to predict whether or not firing would resume tomorrow at midnight. While visiting a Naval base earlier today, Defense Minister Ya'alon said – this war was not over. I am also not willing to hazard a guess on whether or not we will be running for cover come tomorrow – once again. Meanwhile, the community pool in the town of S'derot was crowded today for the first time this summer. The pool has been closed throughout all of the fighting.

I received two e-mails today from the school my son attends – one from the new Superintendent, and the other from the high school Principal. His school is scheduled to begin the academic school year this coming Monday. Both e-mails detailed the psychological services available for students who have been affected by the events of the past month. The radio is also full of programs discussing how to help children who have been negatively affected by sirens and rockets. A study conducted a few years ago concluded that 40% of the children living in S'derot (a town on the Gaza border which has been on the receiving end of rocket fire since 2001) suffer from PTSD. Israelis, myself included, get very frustrated when observers from abroad are dismissive of the Hamas missile fire.

I do not think any major western city has heard the sound of air raid sirens since World War II, (with the exception of Tel Aviv during the Gulf War). Since the start of this current war we in Tel Aviv have heard those ground shaking vibrations, followed by the intercept blasts, with the corresponding sound of the explosions, nearly 40 times. I have no idea what the lasting effects of the events of the last month will be on the children of this country. I am sure it will be less devastating than the effects on the children of Gaza – a tragic reality, compounded by the fact that we did not start this Gaza war.

Even as the possible resumption of hostilities remains, Israel is dealing with the consequences of the war. First, attention has being directed toward the appointment of a Board of Inquiry by the UN Human Rights Commission to investigate possible war crimes committed by the IDF. There was particular anger, followed later by amusement at the appointment of William Shabas to head the committee. Shabas has called for Prime Minister Netanyahu to be hauled before the Court of International Justice, in The Hague, to answer for crimes during Operation Cast Lead. This statement belies both Shabas' "neutrality" and his ignorance. Netanyahu was not in the government in 2008 during Operation Cast Lead. Ehud Olmert was the Prime Minister at that time. Israel plans to take steps to blunt any biased probe of its conduct by creating an independent commission (that will include internationally known jurists) to investigate all of the events that took place in the recent fighting.

Israel is also dealing with the economic consequences of the war. The Defense Ministry has demanded $4 billion increase in its budget – over the next two years – because of the war. Before the war, budgetary discussions in the Knesset primarily focused on how to cut the defense budget. One of the leading items in the news today in Israel was the scene from the Knesset Budget Committee. In one of the unique features of the Israeli political system, that committee has the ability to unilaterally approve the transfer of funds from one ministry to another. Today, when Stav Shafir, (who had been one of the key leaders of Israeli social protest movement 3 years ago, and is now a member of the Knesset Budget committee) objected to the transfer of approximately $800 million to the Defense Ministry from other ministry budgets without any discussion, the chairman of the committee had her forcibly removed from the meeting. The next few days should be anything but boring.

WE CEASE...

HAMAS FIRES.

August 14, 2014

DAY 38 @ WAR WITH HAMAS

Ceasefire Ends – Gets Extended, Maybe

Things are still very fluid, a ceasefire was declared and rockets fell- so things could still change in the coming hour.

Today was a day of waiting in Israel – Would an agreement be reached with Hamas before the ceasefire runs out tonight at midnight? Might the ceasefire be extended, even if an agreement was not reached? You could not go far – anywhere in Israel today – without these questions coming up in conversation. Clearly, Israelis were all hoping this war was over. As the evening proceeded, the picture became clear that no agreement had been reached with between Israel and Hamas in Cairo. The Israeli delegation left Cairo to return home this evening. Hamas scheduled a Press Conference for 9:30 PM. The question remained would Hamas agree to extend the ceasefire, as Egypt was demanding or not? A little after 9:00 PM Hamas delayed its Press Conference. Then, around 9:40 PM, sirens were heard again in Southern Israel. Hamas claimed they were not responsible.

The level of war weariness here is palpable. Israeli TV channels that had been broadcasting special coverage from morning to night during the war, did not interrupt their regular programming tonight. When the missiles were fired, a series of orange rectangles appeared on the screen to indicate the cities that had been targeted. The main radio news channel announced there was a "red alert", and returned to the regularly scheduled basketball game.

The IDF called up more reserve forces today and called back troops that had been given leave. The army made it clear that if the ceasefire is broken, this time, the army would react more vigorously than it did the last time. Was that just a threat? Was the army planning a major action?

At 11:20PM, word began to filter out that the ceasefire had been extended for an additional five days. By all accounts, the Egyptians pressured Hamas (as strongly as they could) to extend the ceasefire. Earlier, the Egyptians had presented their own plan for long term ceasefire, but said it would require more time to fully work out the agreement. At the last moment Hamas agreed to the extension.

It would seem Hamas was not willing to walk away from the Egyptian proposal and return to war – a war, which its people do not want to resume. Tonight. The average Israeli is – on one hand, heaving a sigh of relief, as tomorrow may be another quiet day. On the other hand, Israelis feel frustrated that we all held our breath tonight, questioning whether the war was going to resume or not – knowing there was nothing we could do about the situation.

As midnight neared more rockets were fired at Southern Israel – a total of five. These were not fired in the minutes before the ceasefire was supposed to go into effect, rather during the "ceasefire". Will this ceasefire hold? At this moment it is not clear. Over the past weeks there has been much speculation about the split between the political and military wings of Hamas. Does the continued rocket fire mean that Hamas's political wing is not calling all the shots? The Israeli government has ordered the army to respond to the rocket fire, it's therefore not clear as of now whether the new ceasefire is now in effect.

If the ceasefire does hold, the next deadline is Monday night at midnight. It is unlikely that an agreement between the sides will be released by then – since neither side has been willing to make compromises. Both sides believe they won, and are acting accordingly. Monday night will be another tense night.

August 15, 2014
DAY 39 @ WAR WITH HAMAS
A Confused Nation

We are reaching the end of first day of a five-day ceasefire, a day in which no negotiations have taken place. I just returned from one of the oddest demonstrations I can remember attending. It was a demonstration in support of the Israeli communities situated close to the Gaza Strip;

communities that have been subjected to what is euphemistically called "a drizzle" of rocket fire from Gaza for the last 13 years. The rally held in the center of Tel Aviv at Rabin Square was organized by the residents of the South, with the help of the Tel Aviv municipality.

What was odd about the rally was that neither the speakers, nor the attendees, were really clear about why they were there or what they wanted to occur. Let me be more precise. Everyone did know what they want. Everyone wants peace and security in the communities around Gaza. Everyone wants an end to the current status quo (where it was "ok" to have an occasional barrage of missile fired at southern Kibbutzim and cities. However, verbalizing what must be done to achieve these imperative changes seemed to escape everyone I spoke to amongst the demonstrators tonight.

Batya and Yonatan were typical of those who came from the South to the rally. Both life-long member of Kibbutz K'far Hadarom on the Gaza border, and people whose political views (in normal times) would be considered "Left of center". I asked them what they wanted to happen now. Did they want a broader ground attack? At first they said – No. They believe the only way to solve our situation is through political negotiations. They felt strongly that the government was not doing enough in that area. Yonatan said he had been against a ground action, until the discovery of the tunnels (which he was totally unaware of before this operation). After which, he believed there was no choice.

I shared a conversation with Batya that I had had earlier today with someone who considered herself a lifelong Leftist. I asked if she thought the views of Leftists had changed. Batya thought definitively that the views of Leftists had been changed by the war. She said that it was important to her and her friends to make sure we were doing all we could to achieve a political solution, (something she did not think was the case). However, she continued, they some of the members of the Israeli Left have begun to realize there might be no other choice, but to use force. Yonatan was more skeptical about the use of force. He believes that if only 10% of the Gaza population truly wishes us harm and we could wipe out that 10%, then maybe using force would work. Sadly, he said, that after all of our bitter experiences it has become clear that if we kill one Hamas fighter three come to replace him. Therefore, he feels we must find a political solution.

The same was the case with everyone I encountered at this rally for the South. Everyone I spoke to believed that something has to change. One particular vivacious young woman from *Kibbutz Ein HaSholsha* was holding up a banner that read: "A red alert is not a notice on the TV." (Alluding to the fact that throughout Operation Protective Edge most of the country saw notices on their TV screens when sirens were sounded in any part of the country – but 85% of those TV notices were alerts for very real missile alerts in the communities of the south). I asked her what she thought should be done. She answered frankly she had no idea. She is not in a position of power, but she felt those in power should do something. So I asked her – what if they do not know what do to do? She answered– Then they should resign.

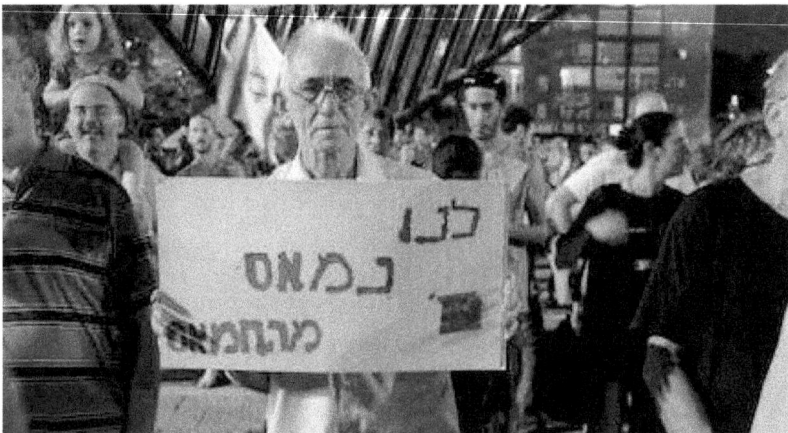

▲ Rally at Rabin Square. Slogan on poster:
"We are sick and tired of Hamas".

A man named Hayim from the southern *Kibbutz Kisufim* was holding up a banner attacking Prime Minister Netanyahu for not doing more. I asked Hayim if he supports a full ground assault on Gaza. He also responded: No. Hayim said he was a member of the Kibbutz community guard. A few months ago they were informed their services were no longer needed. Given the current situation and the recent revelations, Hayim questions how our government could possibly be doing enough if they chose to lower the level of preparedness on the Gaza border just a few months ago?

The responses were the same throughout the rally – whether I spoke to people from the South or from Tel Aviv. Most of the supportive crowd, numbering approximately 20,000, shared the same dilemmas– Knowing things cannot go on as they have been... but having no idea what should be done to achieve change. There was just a resounding consensus that the South must not be forgotten.

The rally in Rabin Square this evening reflects the state of mind of the people of Israel tonight –People are confused. They believe that what has been cannot continue, though they fear it will. Israelis are united in a way we have not been in decades. Israelis believe that they have true enemies, but remain divided as ever, about finding the right resolution to their problems.

August 18, 2014
DAY 42 @ WAR WITH HAMAS
Ceasefire Extended by 24 Hours

At the request of the Egyptians the ceasefire was extended by 24 hours. An agreement was not reached but there are those who are optimistic that another 24 hours could matter. Once again, today was a day when all of Israel held its breath to see what would be decided. The army has ordered that train service to S'derot be stopped (after Hamas showed a video of their clear line of site to the railroad, raising the fear they could fire an anti-tank missile at a rail car. Needless to say, why that problem was not accounted for when that railroad line was built – less than two years ago – raises serious questions.

My son returned to school today. Of course, one of the first things he and his High School classmates did on their first day of school was to participate in a missile drill – i.e. what to do if there was a missile attack. Tomorrow they are going to have a drill on the bus. They were told that in case of a missile attack they are to unbuckle, put their heads between their legs, and get as low as they can. In how many places in the world are students going back to school and these are the first lessons they learn.

For the past few days Israelis have tried to forget that there was a war. Business owners in Tel Aviv who suffered a 50% drop in sales during the war report that business is returning to normal. On Saturday morning I was able to walk again along my favorite place on the beach in North Tel Aviv for the first time, without worrying about a red alert.

After listening to the news media in the last few days the average Israeli could be forgiven if they are confused and do not understand what the immediate future holds. The Egyptian long-term ceasefire proposal of last week – presented as a "take it, or leave it" proposal to both Israel and Hamas – has seemingly been Left behind by both sides. Instead, a less comprehensive proposal is being discussed; one that includes the opening of the border crossings, increasing the distance Gaza fisherman are permitted to fish, and decreasing the no-man's land around the Gaza border. Discussion on all of the long-term issues, such as: disarmament and building a port would be delayed for a month. Under these terms, neither side accomplished very much diplomatically after this war. Israel has achieved very little in terms of limiting the ability of the Hamas to rearm, nor is the agreement regarding the border crossings all that significant for Hamas. In a month negotiations will begin on the larger issues, but there is very little expectation that these will be successful.

There were reports in Israel today that the Israeli Security Services disrupted a Hamas plan to take over the West Bank. That plan called for an uprising that would force Israeli troops to reoccupy the West Bank cities, thus ending the Palestinian Authority's rule. After that Hamas hoped to gain power. The Head of the Palestinian Authority tonight warned that if these reports are correct it would undermine any chance of Palestinian Unity. Whether those reports are fully accurate or not, (and there is no way of really knowing), one thing is clear. While most Israelis have been tuned to events in Gaza this past month, 26 Palestinians were killed by the IDF in the West Bank. Part of that was no doubt caused by the additional violent demonstrations in the West Bank. However, part of these loses are due to the fact that less trained reservists replaced the regular troops in West Bank, who were sent to serve in Gaza.

August 19, 2014
DAY 43 @ WAR WITH HAMAS
The War Resumes

It's amazing how fast things seem to change here. Last night, as I was writing my article, I was trying to fully understand the supposed agreement that had been reached in Cairo. However, at the last moment it become clear that an agreement had not been reached-- though a ceasefire was being extended for 24 hours in order that there would be enough time to finalize the last details. Only overnight did it become clear that none of this was true. It turns out that last night the representatives of the Palestinian Authority were putting forth a patently untrue story as to what Israeli and Hamas had agreed, when neither had. Unfortunately, all of the media (both Israeli and worldwide) reported the story as true. As a result, last night we all went to sleep thinking there had been a 24 hours extension – and that no doubt –over the next few hours an agreement would be reached (for better or worse).

In reality, nothing was further from the truth. By morning it was becoming apparent that neither Israel nor Hamas had agreed to any agreement. Moreover, the sides were far apart. Further complicating the situation, it turns out that the Hamas delegation was split. The political leadership, led by Hamid Mashal seemingly opposed any agreement in which demands like the granting Gaza a seaport were not met.

Any hope that this round of negotiations could be salvaged disappeared late this afternoon when rockets were fired at Be'ersheva. Israel responded by attacking targets in Gaza, and ordered its negotiating team back from Cairo – since Israel had made clear it would not negotiate under fire. As the evening progressed the situation quickly deteriorated with missiles first being fired around Gaza, then at Ashkelon, where a cafe was hit. Next was Ashdod, and finally, toward the Tel Aviv area. The missiles that were fired at Tel Aviv fell in open areas, so there have been no sirens up to this point in Tel Aviv.

Listening to the radio at the moment the rocket warnings seem to be coming ever few moments. It has become almost impossible to hear any radio announcers between the announcement of rocket attacks. Now, we are back to where we where two weeks ago with rocket attacks now coming one after the other. By this evening the Home Front Command ordered all bomb shelters opened from Gaza to Netanya.

Where we go from now is anyone's guess. We are clearly back to square one. If Hamas thinks that Israel will change its positions due to the rockets attacks, they are clearly wrong. They have stated that they are ready for a War of Attrition. The Israeli army has been ordered to go to full war footing and prepare once again for a possible ground attack. It could be that is exactly what Hamas wants to happen; force Israeli into a costly full ground assault on Gaza- They may get their wish, as much as Netanyahu has been against such an assault, if this goes on for very long I do not think he will be able to withstand the public and political pressure to do something drastic.

August 20, 2014

DAY 44 @ WAR WITH HAMAS
Groundhog Day

In many ways, I feel like we are living in Ground Hog's day. Everything in Tel Aviv (and in Israel in general) seems to have turned back five weeks – to the way things were at the beginning of the war. For one week there was silence. For one week television had returned to its normal broadcast schedule. Tonight, my favorite television channel is once again broadcasting from the roof of the city hall (where there is an excellent view of all of Tel Aviv, along with any Iron Dome missile interception that might happen.)

Until now Tel Aviv itself has been spared from any direct attacks and the sirens have not gone off. There was a missile launched toward Tel Aviv a little after 6PM tonight. The intercept was clearly visible from the porch of my apartment. Though had that missile landed it would have hit Tel Aviv, so the red alert sirens were not activated. There seems to be a policy to limit the use of the sirens, (at least in the center of the country), in attempts to limit the psychological effects on residents. Other parts of Israel have not been as fortunate. The South has been living under a constant barrage of missiles. As of 8PM, 170 rockets had been fired on Israel. Israel has responded with a continued series of air attacks on military targets in Gaza.

Economically, however, for business owners, things have gone back to the bad day of the early part of the war. I spoke to Assaf, owner of a restaurant on Dizengoff Street (one of Tel Aviv's main avenues). He confirmed what other business owner had told me – that business was down 50% this July. However, he pointed out that usually business normally grows 100% in July. Therefore, the true loss was 150%. Assaf blamed Prime Minister Netanyahu for not taking more vigorous actions. Whether they blame the Prime Minister or not, tonight most Israelis are confused and concerned. Nobody seems to be providing a clear answer on how, or when, this is going to end.

Netanyahu held a press conference tonight in Tel Aviv. I have rarely seen the Prime Minister as agitated as he was during tonight's conference. He spent an inordinate amount of time attacking his cabinet ministers for publicly criticizing him during the war. Israel has a rather unique political system, where important ministers are often political opponents from other parties (not that of the Prime Minister). During this war we have seen an unprecedented number of verbal assaults of Prime Minister Netanyahu by Foreign Minister Lieberman and by Economics Minister Bennett for not attacking Hamas more vigorously. Even a leading member of Netanyahu's own party, who is also a cabinet minister, felt it was appropriate to attack the Prime Minister tonight – during a live interview, on one of the evening news programs. In the rest of his brief statement and follow-up questions Netanyahu did not go beyond platitudes.

Much of the news media today was caught up with the question of whether or not Israel successfully assassinated Mohammed Deif, the military commander of the Hamas. Deif, who Israel has previously attempted to kill four other times, is the inventor of the suicide bus bombings, as well as a long list of other horrific terrorist methods. As of tonight, it is still not at all clear whether or not Deif is alive. Hamas claims that he is alive. However, tonight there are a number of indications, (including a slip by the Hamas spokesman who briefly referred to Deif as "a martyr", and the fact that the six 1 ton bombs landed on the house where Deif was thought to have been) that the mission was accomplished and Deif was most likely killed. This targeted bombing took place at around 9PM occurred hours after the ceasefire had been broken. Israel hoped that killing Deif would significantly impact Hamas. Many observers question whether that, in fact, will be true.

On a personal level, today was different in a number of ways. The first thing I did when I woke up this morning was to check if there was a text from my son's school – to see if he was going to have a "missile day" (in contrast to the "snow days" he had known in the past). Since there was no such message, he went off to catch the bus to school early this morning. On a very different note, the soldier who I have mentioned a few times in this diary, and who lives with us part of the time, was supposed to get a 4-day leave starting yesterday. By the time he was let out yesterday it was too late to make it to Tel Aviv. Today, as he was heading here, our young friend was informed he needed to report back to the South the following morning. His leave would be over, and once again he may soon be in harms way.

August 21, 2014
DAY 45 @ WAR WITH HAMAS
Light at the End of the Tunnel?

Today in Tel Aviv there is a sense that things have finally changed. There is a feeling that there may actually be an end to this war that will not leave Israel worse off than when it started. I was in the Carmel Market today (Tel Aviv's open air marketplace). Much to my surprise it was packed with people.

The war may have resumed three days ago, but as of now the sirens have not yet gone off in Tel Aviv. Tonight Hamas shot one long-range missile in what seems to have been the general direction of Ben-Gurion Airport. It was way off course, and in any case, it was easily intercepted by the Iron Dome system. Of course, if you live in those parts of Israel closer to Gaza – and within range of their smaller missiles and mortars – the last three days have been days of endless Red Alerts with missiles being fired constantly. Today one person was seriously injured, when a mortar landed next to a kindergarten.

In retrospect, clearly Hamas's decision to resume the fighting, despite having a very minimal number of long-range missiles, will surely be seen as a mistake. Beyond that fact, it has taken a major intelligence break, together with the seeming mistake of Mohamed Deif that allowed Israel to target him – an action that seems to have changed the calculus of events here. It is not totally clear that the assassination attempt was successful. However, what experts on the organization have seen is that Hamas is acting confused. They do not seem to know how to react, which is likely a sign that Deif is either dead or mortally wounded. That confusion explains the fatal miscalculation made by two of the top Hamas leaders last night, by meeting above ground in a place they could be targeting by the IDF. That error resulted in the successful assassination of two of the top leader of the military wing of Hamas, including the person who was in charge of the tunnels, and the person responsible for orchestrating all of the attacks via Egypt on southern Israel. One of the people killed was also someone who oversaw the kidnapping of the Israeli soldier Gilad Shalit eight year ago.

These successful operations will not bring about the immediate collapse of Hamas, but have changed the rules of the game. To have removed almost half the top leadership of Hamas, over the course of just over two days, is a major blow to Hamas. Coming at a time when its rocket supply seems to be reaching the bottom of the barrel is a major blow.

It is hoped that if the Cairo talks get going again, (a step that is assumed will soon follow) that Hamas's first demand will be not for a port, but rather for the targeted killing of its leaders to stop. The sense in Israel is that those talks will not resume until next week. The Egyptians seem to be in no rush to bring the sides back to the table – especially at a time when they believe Hamas is getting what they perceive as "their just deserts" for breaking the ceasefire and ending the talks.

Reports from Qatar today pointed to the fact that Khaled Mashal, political leader of Hamas, told Palestinian President Mahmoud Abbas that he was prepared to have Hamas undertake a War of Attrition for months. Sitting in his hotel in Qatar, Mashal can say that, but the people of Gaza cannot sustain a War of Attrition much longer. Some people in Tel Aviv may be hurting economically (and certainly those who live near the border under constant missile attacks find them much more than a nuisance), but we all have water, electricity, Internet. Our stores are full and our banks are working. This war has its economic costs, and its psychological costs, especially if you live in the South. However, it is clear who can sustain a War of Attrition who cannot. Mashal may think his people can continue, but every report from Gaza says the opposite. We can only hope that the reality will catch up to the political leadership of Hamas, before it's too late for their people.

August 22, 2014

DAY 46 @ WAR WITH HAMAS

Becoming a Manic Depressive

I had almost finished writing an article; of which part of the theme was that for those of us in Tel Aviv it is like there is almost no war going on. This thought ended a few minutes ago when the sirens once again went off in Tel Aviv. Even though we all have confidence in the Iron Dome Missiles, things suddenly seem very different when sirens go off. The reality that we are still at war becomes clear. As my son said, when the sirens go off he can feel his pulse race. I know I certainly would not want to take my blood pressure during those moments. However, the news of the last few minutes underscores how different our lives are then those who live so close to Gaza. As I was rewriting this article, the news broke that a four year old was killed on a Kibbutz near the Gaza border by a mortar that was fired at the same time as the missile was fired on Tel Aviv. The four year old becomes the first Israeli to die since the Hamas broke the ceasefire. But, there's more.

Unfortunately, earlier today an incoming missile hit a synagogue in Ashdod. I must say I am beginning to feel like a manic-depressive. Yesterday, it seemed as if maybe we turned a corner. Yet, now we are back where we started yet again. Of course, neither of these things is true. Yes, Hamas did manage to send a missile that was intercepted over Tel Aviv. But earlier in the war, they were sending four or six missiles at the same time. Now, they have only managed to send one. Of course, for the four year old who will never get to have a life and his parents who will grieve for the rest of their lives, this has been the worse day of the war. All of Israel will mourn the loss of this one four year old.

One of the controversies in the last 24 plus hours, both here in Tel Aviv and on the Internet, has been the attempt by Prime Minister Netanyahu to equate ISIS with Hamas. It is an interesting comparison, especially considering the fact that part of Netanyahu's argument for wanting to keep Hamas in power has been his fear that something worse like ISIS might take over. Netanyahu's claim gained some visual substance today when Hamas circulated photos of their execution of reported collaborators. All 18 supposed collaborators were summarily killed in the streets today. The group of murdered collaborators included at least 2 women. Though the numbers do not come close to those numbers that ISIS has killed, the same blood lust seemed to be in the air among those who were pulling the triggers in Gaza and those who were pulling the triggers in Syria/Iraq.

While they are clearly not the same (and saying they are is an oversimplification,) there are some very clear similarities that should not be ignored. Both movements are religious-based movements, whose political decisions are based on religion. In both cases, those who make decisions believe they are acting in the name of God. In addition, as opposed to some other Islamist movements, both movements believe in the use of violence and both movements use the killing of innocent people as part of their tactics. Hamas generally does not line people up and shoot them like ISIS does. They also did not kill Gilad Shalit because they knew how valuable he was, as opposed to ISIS who used beheading as a tactic. However it was Hamas who invented suicide bus bombings that killed dozens of innocents at the same time. Is Hamas more moderate? You could say that. But Hamas is also run by a generation of men who seem older than the ISIS leaders- people who tend to at least act with more moderation. But are they really that different? I will let you, the reader, decide.

DAY 47 @ WAR WITH HAMAS

Today was day 47. This is now one of longest wars in Israel's history. Only the War of Attrition in 1968-69 and the War of Independence were as long (and though deadly to the soldiers who fought and their families, the War of Attrition took place hundreds of miles from Israel's major cities.) Over 4,000 missiles and mortar shells have fallen inside Israel. This morning, for the first time while the war was going on, I decided to take my favorite Saturday morning walk along a mostly empty beach in Northern Tel Aviv. I had been avoiding this relaxing exercise excursion since there is nowhere to get any cover along that stretch of beach –in case of attack. However, this morning I decided that my mental and physical health were more important than the slight risks involved.

The early morning walk was relatively uneventful, disturbed only by an occasional news flash on my iPhone of an attack close to Gaza. Once again, during most of the day it was the citizens living close to Gaza who were attacked repeatedly. This evening – as has been their pattern – Hamas launched a number of missiles at the center of the country. Tonight was another case when the missiles were not going to strike Tel Aviv, so the sirens did not go off here. While in many ways that creates less fear, it was certainly disconcerting to be walking with my son back from his favorite Pizzeria, with Pizza in hand, and suddenly – without warning – hear a very loud explosion in the air.

There is no question that the attention of most Israeli's was focused today on the Kibbutzim around Gaza. The death of the 4-year-old Daniel Tragerman from a mortar shell yesterday underscored the price the people in those settlements have been paying for years. In this recent war the level of fire on the settlement has increased exponentially, and especially in this newest stage. After the ceasefire that failed, an ever-higher percentage of fire has been aimed at those targets that are in short range of Gaza. Since there is no more than a 15-second warning before an attack on those locations, there is little the residents can do.

10 days ago I had the chance to meet a large number of residents from the South. I wrote a story about them coming to Tel Aviv to demonstrate and demand a permanent solution. I cannot help but reflect on those brave people, who all they want – after all these years of shelling – is for it to end. The words of the eulogy Moshe Dayan (then the Defense Minister) gave in April 1956 for Roi Rotenberg, the security coordinator of *Nachal Oz* who was killed by terrorist, the same Kibbutz where Daniel Tregerman was killed yesterday, are stuck in my head. "We have no choice but to fight, declared Dayan. "This is our life choice," he said, "to be prepared and armed, strong and determined, lest the sword be stricken from our fist and our lives cut down."

Dayan went on to say "We are a generation that settles the land, and without the steel helmet and the cannon's fire we will not be able to plant a tree and build a home, Let us not be deterred from seeing the loathing that is inflaming and filling the lives of the hundreds of thousands of Arabs who live around us. Let us not avert our eyes lest our arms weaken." Dayan also added in the memorable eulogy: "Let us not cast the blame on the murderers today. Why should we deplore their burning hatred for us? For eight years they have been sitting in the refugee camps in Gaza, and before their eyes we have been transforming the lands and the villages, where they and their fathers dwelt, into our estate." It had now been not eight years, but 66 years and the hatred has no doubt grown stronger and not weakened.

Tonight Egypt invited the sides to permanent ceasefire talks in Cairo. In Israel the sense is that Hamas is not ready to change its position, and thus, will not enter into any permanent ceasefire arrangement at this moment (of course that could change.) Tonight Israel destroyed a 14-story apartment building in downtown Gaza. Military sources claimed parts of the building were being used by Hamas for command and control. Until now, that part of Gaza has been largely untouched by Israeli aircraft. It would seem that now Israel is increasing the pressure. There is a growing sense – with school scheduled to start for most Israeli school children on September 1st – that if Hamas does not change its positions this week, Israel will use ground troops, once again, to try to put additional pressure on Hamas. Is that a mere threat or is it real? Time will tell. Tonight, at around 10:30 pm, missile alerts went off in Northern Israel, far from Gaza. It seems that Palestinians in Lebanon fired a Katyusha rocket at Israel.

Finally, tonight on the news they devoted an entire segment to covering the psychological effects on a country at war – e.g. the large number of premature births, the 66% increase in the use of tranquilizers, and a variety of the other consequential side effects. In reality, how much has this war affected the average Israeli? This is much less traumatic than the events of the second Intifada, when buses and cafes were being blown up. Thankfully, our casualties have been very light – except for those who live around the Gaza Strip, we all have sufficient warning if a rocket is headed our way to take precautions, and beyond that, we an anti-missile system that works. However, what has been hurt is our sense that we have become a normal country. There was that sense in the last few year of normalcy, that Tel Aviv was an extension of Silicon Valley just with a nice beach. We still have a nice beach and still have more App developers per square mile then anywhere else but Silicon Valley, but one thing we all realize once again this is not a normal country, this is a country that for whatever reason has some neighbors who would like us all dead.

August 24, 2014

DAY 48 @ WAR WITH HAMAS

War with Hamas

Today was day 48 of the war. I remember by the second week of the war a number of the Israeli correspondents saying we had reached the ink time of the war (when nothing could go well and things could only go wrong) and thus, it was time to end this war. They would never believe that four weeks later the war is still going on. As wars go it was a rather uneventful day in Israel. The most significant event was the Palestinians shelling the *Erez* crossing – the place where they are critically injured and chronically ill travel to Israeli hospitals for treatment on a regular basis. Hamas managed to seriously injure a number of cab drivers, one from East Jerusalem and the others, Bedouins from the Negev who were waiting to transport the sick for care. If that was not enough, they also targeted the *Kerem Shalom* crossing, the passage through which Israel supplies food for the residents of Gaza on a daily basis. My son had the most insightful view of their actions — suggesting that Hamas wants to force Israel to close the crossing so then they can charge Israel with starving the people of Gaza.

It looks like both sides are getting ready for a long War of Attrition. Today Prime Minister Netanyahu stated Israelis should be ready for a war that lasts into the new school year (which is scheduled to start in a week. As I have said before, it is clear which side is best suited to hold out, however difficult it would be. It will be easier to hold out than to suffer the dead soldiers that would result in forcing an end, by reconquering Gaza with ground troops. During the second Intifada the Israeli people were able to manage for a long time — as long as our death toll remained less than the number of usually killed in car accidents. At the moment, the Israeli casualties are thankfully even close to those numbers. Clearly, if this conflict continues we might make more limited ground raids, (especially aimed at the sites of the mortars). However, who knows ... as unpredictable as the entire situation has been in the past few weeks, and the extent to which the Israeli government has been wrong in predicting what Hamas might do, this war could suddenly end tomorrow, or go on for another month.

Today was the profoundly tragic funeral of Daniel Tragerman, the four-year-old boy who was killed Friday afternoon. Gideon Levy wrote an article in today *Ha'Aretz* about the fact that we know nothing about the 478 children in Gaza who have died in this war. He is right. I, for one, certainly feel for them, but they are not my primary responsibility. Steve Wozniak who is visiting Israel said it all while visiting in Jerusalem and the Gaza Strip earlier today: "If Israel did not react, the rockets would continue anyway. If Hamas halted rockets, Israel would not attack them. Peace."

August 25, 2014
DAY 49 @ WAR WITH HAMAS
Seven Weeks of War with Hamas

It has been another day of something between war and peace in Tel Aviv. This war began seven weeks ago tonight. Yesterday came and went without any sirens on Tel Aviv. One missile was fired in our general direction, but it fell in an open area. Last night my daughter (who finished her army service earlier in the year) returned after four months abroad. We all hoped that she would bring the good luck that would end the war. As a matter of fact, as I was waiting for her to emerge from customs, in the wee hours of the morning, my twitter feed became full of Palestinian sources claiming that a ceasefire agreement was imminent. Of course these seemed to be the same sources claiming that an agreement was on verge of being signed last week.

This morning at around 8:00 am alert sirens went off twice in Tel Aviv. They both turned out to be just a false alarm. This evening, as I was writing this piece, Hamas tried firing in the general direction of Ben-Gurion Airport, which they seem to have missed. That rocket was heading toward the suburbs of Tel Aviv. Thankfully, that rocket was intercepted by Iron Dome. Since this is a small country, I could hear the missile interception from my bedroom window in downtown Tel Aviv. Tonight another missile fell on Northern Israel from Lebanon. When it happened a few days ago Israel decided not to retaliate leaving to the Lebanese, since they were not effective in stopping the new rocket fire, Israel has responded with artillery fire on Lebanon.

The people in Tel Aviv went about their business today. They have long internalized the realities of having to possibly run for cover. However, as the number of missiles fired keeps decreasing, the intermittent need to quickly find shelter seems to have become – slowly, but surely – a minor nuisance in life. Needless to say, what is a minor nuisance in Tel Aviv, dominates the life of those who live near Gaza. As Hamas has been running low on its long-range missiles, and even depleting its reserve of medium range missiles, it is turning more and more to launching its short-range mortars.

Hamas has been firing on the settlements around Gaza in larger and larger numbers. As of 9 PM this evening, Hamas launched 130 rockets Israel today – the overwhelming majority of these missiles landed in the area immediately surrounding Gaza. Most of the Israelis living near Gaza have decided to take extended vacations until the fighting ends. While there has been some controversy regarding this issue, almost all of the country understands the decisions of these fellow Israelis. If anything, there has been anger at the government for not doing a better job providing for these residents. Israel knows where most of the mortars are located. Many are located in schools and refugee camps. Tonight, Israel warned those currently in the school in which the mortars that killed Daniel Tregerman were located to leave the building (since it is planning to bomb the mortar stockpile.)

Israel has been engaging in what appears to be a contradictory policy in the past few days. On one hand, it has been more aggressive in its attacks on Hamas targets. It seems to be doing its best – successfully – to limit civilian deaths. This translates into fewer attacks, but attacks on higher value targets (including more attacks on the leadership of Hamas and the other military organizations in Gaza.) One the other hand, with this strategy the chances of something going wrong are high, and Hamas is hoping that Israel will accidentally kill a large number of civilians.

There continue to be reports of a ceasefire being imminent. The Egyptian plan calls for a one-month ceasefire, followed by discussions on the major issues – e.g. a port and disarmament. So far, everyone but Hamas seems to have accepted the plan. As of now, Hamas has not insisted on guarantees that its demands on issue such as the port be met up front. It is very hard to analyze what chances exist for a successful ceasefire. From the very beginning of this offensive, the accepted wisdom in Israel was that Hamas did not want this war, and as such, that it would accept the first ceasefire ... then, the second ceasefire ... then, the third ceasefire. Will Hamas accept this one? …

One clear victim of this war has been the popularity of Prime Minister Netanyahu. President Obama can no longer be jealous, as he seemed in the interview with Jeffery Goldberg with regard to his counterpart's popularity. On July 23rd 2014, 82% of those polled were satisfied with the performance of Prime Minister Netanyahu. Tonight that figure stands at 38%, representing a 44% drop in one month. Netanyahu joins the long list of Israeli Prime Minsters whose popularity was very high when a war began, and plummeted by the end.

August 26, 2014
DAY 50 @ WAR WITH HAMAS
Ceasefire...
Theoretically, a ceasefire went into effect at 7 AM. I say "theoretically", since in the first few minutes of the alleged ceasefire there was non-stop missile fire from Gaza on Israel – starting with barrages of missiles that slowly died off. Later in the evening some Hamas leaders came out of their underground bunkers, which is a sign that this may actually be a real ceasefire. The new ceasefire has no expiration date. It is effectively almost identical to the first ceasefire agreement that was offered to the Hamas as the war began. Under the terms of the agreement, the border

crossings (especially that with Egypt) will be open to bring in supplies. One month from now, negotiations are scheduled to take place on all of the open items (such as: disarmament, ports, airports and whatever else.) There is no question that during the coming month Hamas will try to produce new missiles. Thus, if war does resume in one month Hamas will have made strides toward replenishing their supply of missiles.

The Israeli security cabinet had not approved this current agreement between Israel and Hamas. Netanyahu did not have a majority in the cabinet in favor of the agreement (with neither his right- wing cabinet members – or possibly even Tzipi Livni being in favor). However, Netanyahu went and agreed without their approval.

Tonight, both sides are claiming victory. There will be much to analyze in the coming weeks. What seems clear, for the moment, is that Hamas gained nothing. After insisting throughout the war that it would not agree to a ceasefire without getting a port, Hamas seems to have agreed to a ceasefire merely based on the promise of later discussion on the matter. It would seem that the state of Gaza and their dwindling reserves of missies convinced Hamas to agree to what it chose not to accept before there were 2,000 dead in Gaza.

Of course, this is the 5th major ceasefire called in this war. Will it hold? No one knows for sure. The ceasefire may hold for a few months. There is nothing in this agreement to ensure this situation will not return in a couple of months, or a year or two. Nothing significant has changed. When all is said and done, there will be almost 2 million people living in a small area with no real means of support, and no real hope for the future – but that is a discussion for another day.

This day itself was a difficult one in Israel. In Tel Aviv, the day began with a missile assault at 6:30AM in the morning. While the missile targeting Tel Aviv was intercepted, one missile headed for Tel Aviv failed during the launch phase. Instead of reaching Tel Aviv, the missile fell on Ashkelon. Since the missile was not expected to land in Ashkelon, it was not intercepted. That missile caused significant damage in Ashkelon and also wounded 50 people. The missile fire continued all day. One missile landed in the playground of a Kindergarten in Ashdod. Fortunately, the school was closed.

In the hour just before the ceasefire, major attacks were made on the communities around Gaza. In one of these attacks a 50-year-old Israeli was killed and three Israelis were seriously wounded. If this is indeed the end then Hamas fired a total of 4594 missiles on Israel, 2532 fell in Israel in open areas and 116 in built up areas, 188 fell in the Gaza Strip, 735 were intercepted.

If this is indeed the end of the war, numerous questions remain. Many questions will be asked in the coming days and weeks. Tonight in Israel there is what I would call "skeptical optimism". It looks like Hamas has very few remaining missiles and a population that is tired of war. As a result, Israelis are hoping that today will indeed be the end of this war. On Monday the school children of Israel are scheduled to return to school. Israelis hope it will be into a period of peace.

The Aftermath

The war did indeed end on the 50th day. While all the previous ceasefires were temporary by design (and momentary in fact), the current ceasefire is holding (as of this writing). At this time, there are no indications of when the fighting might resume. The war is over, for the moment, but left many questions that deserve to be examined. For some questions, it is too soon to formulate an answer; others may never be answered. To the families – on both sides – who lost what is dearest to them, there will never be a sufficient answer to explain or justify their losses. For all those – on both sides whose lives were cut short because of this war, there will never be ample justification for why they died.

The first question that needs to be asked must be: Was this war avoidable, or phrased differently, why was this war fought? First and foremost, it must be said that this was not an accidental war; nor was this war the result of a series of events that escalated out of control. Hamas deliberately started this war. Before the hostilities were fully underway, the Israeli government was explicit in saying that it was not interested in pursuing confrontation. In addition, Israel clearly stated it would refrain from anything but a symbolic response, if Hamas ceased attacking Israel. Instead of terminating its attacks on Israel, Hamas expanded them. Hamas undoubtedly wanted a confrontation with Israel.

That still does not answer the question of why? The war seems to have caught Israel by surprise. No one in Israel predicted the possibility of a war with Hamas in Gaza during the summer of 2014. The conventional wisdom held that Hamas would not want a war; they were too weak, and too afraid of losing control of Gaza. Yet, all the "conventional wisdom" turned out to be wrong. It appears that Hamas decided it had nothing to lose from inciting a war with Israel. The calculation was simple. Since Hamas's status quo was unsustainable, they hoped by starting the war, they could reshuffle the deck enough to be able to improve their situation. It is unclear whether Hamas set any real goals for the war.

Hamas might have thought they could overcome Israel's Iron Dome defense by overwhelming it with too many missiles at one time.

War could have seemed like a potentially effective strategy. It became crystal clear early on, however, that the Iron Dome system was eminently capable of successfully dealing with a multitude of simultaneously launched missiles.

Hamas also seemed to believe that its secret tunnels would provide them with a significant advantage in a war with Israel. There was a fundamental fallacy in that approach. If tunnels were going to be used to attack Israel, then do it in a time of relative peace – and not in the middle of a war, when there are tens of thousands of troops on alert in the area. Israeli troops were indeed a target for the Hamas forces when Hamas fighters exited the tunnels. But the presence of so many heavily armed combat soldiers ensured that while any attack might prove to be a short-term tactical success, the Israeli response would put a swift end to the attack.

So, to return to the original question – Could this war have been avoided? The preliminary answer has to be no, since it was Hamas who decided to start the war, and the only way Hamas might have chosen *not* to start this war was if its fundamental status had been better. Hamas' problems were primarily not with Israel. Hamas' difficulties were with Egypt, who had been closing Hamas' tunnels and keeping the Rafiah crossing shut most of the time. From the time the Muslim Brotherhood fell from power in Egypt, Hamas was seen as the enemy by the new Egyptian government.

Thirty-one Egyptian soldiers were killed in Sinai on October 24th, 2014. The Egyptian government saw Hamas as complicit in the killings, and has since taken every action it can against Hamas. It has created a one- kilometer wide buffer zone between Gaza and Egypt thereby ending all smuggling operations. The Egyptians have also completely closed the border. Hamas and the people of Gaza are paying a heavy price, and even if Israel would like to change the circumstances, it is without the power to do so.

Military Lessons from the War

There are a number of military lessons to be learned from this war with Hamas. First, and probably of paramount importance, Israel's investment in anti-missile technology has proven itself worthwhile. Despite the naysayers, Iron Dome performed beyond all expectation. The total overall intercept rate was 84% of all fired missiles fired to land on populated areas. This is high, but clearly not even close to 100%. The number of Iron Dome intercepts reached close to 100% of missiles fired at central Israel. The lower intercept numbers are no doubt reflective of the fact that the closer to the Gaza border one gets, where the window to intercept is under 20 seconds, the lower the success rate. In the central area of the country, where there is a two minute period between the firing of a missile and its impact, and where a decision was made to fire interceptors at every incoming rocket, the IDF achieved nearly 100% success intercepting and destroying Hamas' rockets.

The one major exception was the missile that landed in Or Yehuda, near Ben-Gurion airport and resulted in many foreign airlines temporarily canceling their flights both into and out of Israel. The decision not to fire at the missile was intentional. The determination was made by an Air Force officer who was concerned that any intercept directly over the area of Ben-Gurion airport might prove problematic. There has been much discussion about huge gap between the cost of each intercept, compared to the cost of each missile fired at Israel by Hamas. That topic has always been a red herring. Even if the cost of an interception is 10, or even 20 times the cost of the weapon fired by Hamas, Israel's GNP last year was $289 billion dollars, while the GNP of Gaza was probably less than 1% of Israel's figure. So, who can better afford to keep firing missiles and interceptors? For Israel, even if you were to exclude the cost of loss of human life, and the concomitant political pressures it would cause, a simple economic calculation with Israeli apartments (even in the areas around the Gaza Strip) costing hundreds of thousands of dollars indicate that every time a Hamas missile was intercepted, money was saved.

The one military threat for which Israel was not prepared was short-range mortar fire, which consequently killed a number of Israelis during the war. Early on, the lessons of previous wars, still unlearned, brought troops too close to the border, leaving them exposed to mortar fire. There is now a technical solution to combat mortar attacks. Rafael, the same company responsible for Iron Dome, has already tested such a system. It will take two years before that system is operational.

Another military lesson of this war seems relates to the continually insufficient investment in Israel's ground forces. In three out of the four wars that Israel has fought in the last decade, it has been the infantry forces that have been forced to carry the largest burden in the fight. Yet, we continue to limit investment in those forces. The tragedy of the APC, in which seven fighters were killed in an antiquated vehicle that a previous commission had recommended never be used in combat. To say the bitter and tragic outcome was egregious is a gross under-investment. Furthermore, the average combat ground force unit (with the exception of several elite units) do not have the type of modern navigation and communication equipment found on the average civilian for the better part of the past decade. A relatively modest investment in equipment and training would go a long way to increasing the effectiveness and power of the ground forces, though all of them performed well.

The Limits of Air Power

Once again, the Gaza war showed both the effectiveness and the limits of offensive air power. Despite operating in a near risk-free environment, the Air Force was able to degrade, but not eliminate the firing of rockets on Israel. Since the 1920s, Air Forces around the world have been over-promising what they are able to accomplish. The IAF is no exception. It was able to achieve its operational missions successfully. However, that did not always translate into an achievement on the political front. While it successfully destroyed many of the missile launchers and missile production facilities, the IAF was not able to stop the firing of missiles on Israel, regardless of the number of missions it flew.

The Air Force was limited by the overall limitation on Israeli military power – the fear of injuring civilians. Of course, this is part of the larger problem of asymmetric warfare that Israel has been facing since the Yom Kippur War. Throughout the war, the Air Force tried mightily to limit the collateral damage, and was relatively successful in achieving that goal. Near the very end of the war, when the decision was made to destroy a number of the high-rise apartment buildings in Gaza City, the Air Force carried out that mission very efficiently, destroying the targeted buildings without causing casualties. The destruction of these symbols of Gaza's potential better future might have been a major impetus for Hamas finally agreeing to a ceasefire.

The Tunnels

One of the ongoing controversies surrounding the war has been the issue of the attack tunnels dug by Hamas that terminated inside Israel. The fact that Hamas had created these tunnels was well known. They had revealed the existence of the tunnels as far back as 2006 (when a tunnel was used to kidnap Gilad Shalit.) In the weeks leading up to this recent war, the potential threat posed by the tunnels was the subject of extensive intelligence initiatives. Despite those efforts, the tunnels were not considered a truly strategic threat. Attempts were repeatedly made to try to find and destroy any tunnel that reached Israel, but with little success. There was clear intelligence regarding a tunnel that existed in the Southern part of the Gaza border, in the area of *Kerem Shalom*. However, all attempts to locate the tunnel failed.

As the war began, the tunnels were largely seen as just one of many threats – certainly not a danger at the same order of magnitude as were missiles. During the early days of the war, the realization of the perils posed by the tunnels gradually increased. Immediately after its rejection of the latest ceasefire proposal, Hamas staged another attack. Everything changed on the morning of July 17th, when 17 Hamas fighters emerged from a tunnel near Kibbutz Sufa. The emergence of these fighters was captured on video by an I.D.F. drone that had been flying in the area. The film was widely shared. It now dawned on the I.D.F., the Cabinet, the residents of the areas around Gaza, and the wider Israeli public the overwhelming threat the tunnels actually presented. As a result, the cabinet finally authorized a ground assault on Gaza to eliminate the threat of the tunnels.

It turned out that the I.D.F. was ill equipped to destroy the tunnel quickly. In one of the most celebrated gaffes of the war, Defense Minister Ya'alon, claimed that it would take Israel 2 to 3 days to destroy the tunnels. It in fact took nearly three weeks. The Engineering Corps soldiers were not trained for the task; there was not enough equipment to do the job; there was altogether not enough of the proper equipment altogether. The I.D.F. was clearly not prepared for the job of destroying 30 well-built tunnels. Of course, the I.D.F. did what it does best – it improvised. Instead of using special explosives designed to destroy tunnels, the I.D.F. Engineering corps used strings of land mines. They drafted civilian drilling equipment to join the limited number of military rigs available. All of this took time, time when Israeli soldiers were mostly static inside Gaza, providing inviting targets for Hamas snipers. In the end, the job was completed and all of the tunnels were destroyed. However, it is clear that in any future war the I.D.F. Engineering Corps will have to be better prepared.

Limited Use of Reservists

Another tactical aspect of the war should be noted, and that is the decision to deploy almost exclusively, active duty units to the ground assault on Gaza. Reservists were called up; largely to replace the regular enlisted soldiers guarding Israel's other borders. This represents a significant departure from previous wars, which were fought largely by reservists. Operationally, the decision seems to have worked out well. The regular army units were better trained, and, by-and-large, better equipped to fight. It also served to lessen any criticism of the operation (something that reservists have always done and have been freer to do so than regular army personal. The long-term implications of this decision remain unclear.

An overarching lesson from the war is that Israel has become better at defense then offense. Thanks to Iron Dome and protective measures taken in the past decade all around the Gaza periphery, casualties and damage was extremely limited. Israel was extremely reluctant to deploy its overwhelming offensive force, as that would have resulted in a large number of civilian casualties as well as casualties among Israeli forces. In the end, the defensive capabilities of the IDF gave it the luxury to limit its offensive actions. There can be no doubt that in the coming years, our defensive capabilities will only increase.

The Economic Implications

The Gaza War's impact on the Israeli economy varied greatly by sector, and of course by location. The one area that suffered the most from the war was tourism. Incoming tourism plummeted, as it had done during previous wars (and of course during the second Intifada.) Hotels, restaurants and other businesses dependent on the patronage of tourists suffered the most. Restaurants experienced the most severe decline in business – suffering a loss of both tourist and local Israeli diners (who refrained from going out to eat during the war.) Some restaurants made up for their sit-in loss of domestic business by the increase in take-out orders, but that still did not make up for the losses. The government agreed to compensate businesses located up to 40 km from the Gaza border (the reach of Hamas' medium-range missiles.) Businesses located in the Tel Aviv metro area did not receive compensation.

The total cost of the war will obviously have an impact on the Israeli economy. The budget deficit next year will be higher than expected. However, the direct costs of the war were covered by the reserves already included in the Israeli budget. At the end of November, Finch, (the international rating agency,) lowered Israel's forward forecast from positive to neutral, citing the war and the ongoing uncertainty of the security situation.

Israel & The World

The Gaza war, once again, underscored the constraints of asymmetric warfare in the age of instant communication. Initially, Israel benefited from widespread international support for its campaign against Gaza. There had been some criticism against Israel's policies vis à vis Gaza and the Palestinians. Though, in general, since it was patently clear to all that Hamas attacked Israel, most countries respected Israel's right to respond and defend itself. As the war continued, however, the images on the television screen began to negatively impact world opinion.

The images from Israeli were boring and not very demonstrative of the psychological trauma caused by the events of the day. How many times can you show Israelis running for shelter; watch a rocket streak out into the sky to intercept the incoming missile; and then watch as Israelis seemingly return to their normal lives. The first time, it is possible the images will move you. The second and third time, you may think they are interesting photographs, or videos – but after that …

On the other hand, watching day after day, as Israel responded to Hamas attacks with bombs and missiles that would consistently hit their targets, bring about real destruction, and often, albeit by accident, cause the deaths of innocents, the world's responses were different. The world press could not film the deaths and devastation in Syria, so instead, all of the coverage moved to Gaza. As a result, the world's frustration at its inability to stop the killing in Syria (and in other conflicted regions around the world) seemed to be focused at Israel. Day-in and day-out worldwide media coverage centered on the deaths in Gaza. The fact that missiles were being fired continually at Tel Aviv was sometimes mentioned, almost always as an afterthought.

There was almost never a report mentioning the fact Israel was prepared to end its attack at any time – providing that Hamas stopped firing missiles at Israelis cities. Reports were also remiss in not highlighting the fact the Israeli government had accepted every ceasefire proposals, only to see each one violated by Hamas. Most of the world's governments understood that fact. Most even understood the sad reality that civilian deaths – however terrible – were the inevitable outcome of war. At the same time, the citizenry of many of the same countries (with the exception of the United States) who only saw the images of destruction and death in Gaza came to the conclusion that Israel was to blame.

Israel & World Jewry

This war was both similar, and at the same time very different, from other wars when it came to the reactions of Jewish communities throughout the world and how those views correlated with the views of the Israeli public. As in all wars, there were the usual "Emergency Support for Israel" campaigns. Wars are always a great vehicle to raise money, and the United Jewish Appeal (UJA), Jewish National Fund (JNF) and other groups all took advantage of the war to do so. The mainstream Jewish organizations articulated all of the usual platitudes of encouragement and solidarity. What seemed rather different this time was the gulf that developed between average Israelis and those Jews around the world (especially in the United States) who consider themselves "Left-of-center Zionists", (e.g. groups like J Street or self-professed intellectuals like Peter Beinart).

In an earlier section of the book containing my Tel Aviv Diary chronicle, you can sense the frustration I felt when I wrote about this disconnect during the war. There was a very real feeling in Israel that this was a just war; a war we had to fight. It is possible that conclusion was drawn from raw emotion intensified by living an entire summer under missile fire. This feeling profoundly influenced all but the most extreme Left-wingers here.

For the duration of the summer, the sins of occupation were forgotten. All that mattered to us was finding a means to stop the rockets. The war revealed a true rift that exists between most people living in Israel and most of those who do not. American Jews stopped visiting during the second intifada and the second Lebanon war when rockets were landing in the North, as well as during this war, when rockets were landing all over. (Although it is interesting to highlight that almost all of the Americans already here when the war began chose to remain in the country until the originally scheduled end of their trips). This time, there was a sense that too many of our "friends abroad" were excessively concerned about what was happening in Gaza, with an overall disregard for what was happening in Tel Aviv (not to mention Ashkelon and S'derot.)

Perhaps the incident that best symbolized the Diaspora-Israel divide was a solidarity rally that was held in Boston. The rally, sponsored by the Greater Boston Jewish Federation, originally included J Street as a co-sponsor. The day before the rally J Street pulled back its sponsorship. As it was reported in *Ha'Aretz,* the spokesman for J Street wrote:

> I initially accepted your invitation fully, aware that as you said, this would not be 'a J Street rally.' I agreed to participate on the basis of your assurances that our movement's voice would be represented alongside those of others in our community. At the outset, you asked me to suggest potential speakers and invited my input, in shaping the tone and content that would be inclusive of J Street. Even though I was disappointed that the roster of speakers did not include a pro-Israel, pro-peace

perspective, and that the feedback you solicited from me was barely reflected in the rally's messaging points, I appreciated the difficulty of the task you had taken on … What was missing for us in this rally, and what ultimately precluded our co-sponsorship, was that despite our efforts, there was no space made to raise the issues that follow from our commitment to Israel's Jewish and democratic future. There was no voice for our concerns about the loss of human life on both sides, or the acknowledgement of the conflict's complexity and that the only way to truly end it is through a political solution."[7]

Distilling the statement above, you conclude that for J Street, and others, one cannot support Israel without talking about Palestinian suffering. For Israelis, even those definitively on the Left, there are times which one should be supportive of the government (especially a government that seemed to be acting as carefully as the Netanyahu government was during this war) without qualifying it with understanding for the other side.

Domestic Dissent

The true test of a democracy is how it fares during a war. This is an oft-repeated statement, but one that highlights an assignment that most democracies have failed dramatically over the years. All one has to do is think about what the United States did to Japanese-Americans during World War II or how the US acted during the early years of the Cold War toward those believed to be Communists, to prove that democracies often perform very poorly in defending the rights of those who oppose their accepted philosophies. In the United States, that situation slowly changed during the Vietnam War. As the war became more controversial, protesting the war became more and more accepted. Israel does not have a strong tradition of anti-war protests. After the Yom Kippur War, there were demonstrations. However, these protests were not against war, per se; they were directed against the government who was held responsible for the results of that particular war. One notable exception was the very controversial first Lebanon War.

During the recent war, *Tzuk Eitan*, there was broad national consensus. Still, there were those who expressed opposition – some to the war and more to the numbers of casualties inflicted on the Palestinians. The anti-war demonstrations were met with counter-demonstrations. At times, the situation became ugly. But, it is interesting to note that after a few days of confrontations in Tel Aviv, both sides decided to meet together in Rabin Square for a rather civilized dialogue – one that did not change anyone's mind, but did manage to cool some of the anger each side felt toward the other.

The animus seen here this past summer is not totally unique or new. Israel has had to deal with an ongoing intolerance to opposition regarding government policies for over 30 years. In February 1983, Peace activist Emil Grunzweig was killed by a right-wing protestor at a Peace Now demonstration. Thirty years later, while it would be hard to say the problem has become worse, (since there is nothing worse than having a demonstrator murdered by an opponent), the situation has not gotten better. The pervasive perception is that though no one else has been killed for voicing an unpopular view, the dissonance continues to escalate to new heights. In October, President Rivlin spoke at a conference on the future of democracy and tolerance in Israel, stating: "I'm not asking if they've forgotten how to be Jews, but if they've forgotten how to be decent human beings. Have they forgotten how to converse?"[8] It remains a major challenge to civil Israeli society to find ways to ensure dissent is accepted – even when it is least appreciated.

Effects of the 2014 Gaza War on Israel

At the time of this writing, almost four months have passed since the end of the Gaza War, and still every time a motorcycle goes down our block making a sound very similar to the sound of the sirens going off, I tense. I have lived in Israel during wars. I have been in the army and several times encountered situations that were dangerous. However, back then I was both young and without a house full of family and friends. As an adult with family responsibilities there was something deeply disturbing about never knowing when sirens would go off; when a missile would be heading your way. Logically, it was very easy to say that "we have Iron Dome and nothing will happen". And, there is no doubt that Iron Dome made all the difference. During previous wars, the population often moved out of affected areas and took extended 'vacations' away from the action.

This time, no one took vacation. No one left their house unless absolutely necessary. Intellectually, we knew we were safe, but that fact did not lessen the psychological impact of this war. What the longer-term impact of the war will be is harder to fathom. What lasting affect will *Tzuk Eitan* ultimately have on Israeli schoolchildren — who lost their summer, and are not as hardened as we adults are to the realities of living here? I cannot forget the interview I conducted with a 20-something-year-old, from a Left-wing Kibbutz, who said to me, "this has not changed my view of the need for peace, but I realize — maybe for the first time — that there are people out there who really want to kill me."

WHAT DID THE IDF ACCOMPLISH IN GAZA?

DEMOLISHED 32 HAMAS TUNNELS LEADING INTO ISRAEL

DESTROYED 60-70% OF HAMAS' ROCKET ARSENAL

ELIMINATED OVER 1000 TERRORISTS

OPERATION PROTECTIVE EDGE SEVERELY WEAKENED HAMAS. ALL OF ISRAEL IS NOW SAFER.

Final Conclusions

It is too early to offer a true history of the nasty little war fought this summer. It is too early to know what the short- and long-term impact, the consequences, will be. For a Middle East in total turmoil, the events of summer 2014 may end up being just a footnote – except for those who lost family members and other dear ones. For the moment, what can be said is that this war has deepened the hatred and distrust on both sides of the Israeli-Palestinian conflict. This war has pushed the hope of peace even further away. *Tzuk Eitan* was one of the longest wars in Israel's history – 50 days. Yet, despite all of the problems mentioned throughout this book, neither Israeli society, nor the Israeli economy collapsed.

Israel persevered through it all – as it has endured many other trying events in its history. For some time, there has been a misunderstanding of Israel throughout the Arab world. As Defense Minister Ya'alon recently stated: "the spider web issue is no longer valid". Ya'alon was referring to a speech given by Hassan Nasrallah in May 2000, two days after the Israeli withdrawal from Lebanon. In that speech, Nasrallah claimed that Israel was weaker than a spider web. Israelis have endured much since that speech – the second intifada with all of the bombings, the second Lebanon War when most of the North was targeted, and now a second war from Gaza where Tel Aviv has been repeatedly, but unsuccessfully, attacked.

Despite all that was happening this summer, new immigrants kept coming. Tourists might have stayed home, but those who decided they wanted to spend their lives here concluded that, war notwithstanding, Israel was a better place than from where they had come. The war has left its scars, as all wars do. Most sadly, the war needlessly and mercilessly ended the lives of many who should have had a chance to live out their lives. In the end, it will likely prove to be just another brief chapter in the history of Israel. The killing will go on, and Israelis will continue to persevere. –M.S.

ENDNOTES

1   Israel: A Personal History, p. 447.

2   Bloomberg View, March 2, 2014, Jeffery Goldberg, http://www.
     bloombergview.com/articles/2014-03-02/obama-to-israel-time-is-running-out

3   Ha'Aretz, March 5, 2014, Barak Ravid http://www.haaretz.com/news/
     diplomacy-defense/.premium-1.577958

4   Ravid, "Livni: Palestinian prisoners will not be released, unless a framework
     deal is reached Ha'Aretz, March 18th, 2014, http://www.haaretz.com/news/
     diplomacy-defense/.premium-1.580496

5   Text of President Obama's Press Conference with President Park of the
     Republic of Korea, April 25, 2014, (White House Press Office).

6   Ha'Aretz, Ori Kashti, June 2, 2014, World B'nei Akiva Chief calls for price
     of "blood" for Israeli teens' murder.

7   J-Street pulls sponsorship from pro-Israel rally in Boston, by Alison Kaplan-
     Sommer, Ha'Aretz, July 21, 2014, http://www.haaretz.com/blogs/routine-
     emergencies/

8   http://www.jpost.com/Israel-News/Politics-And-Diplomacy/President-
     Rivlin-Time-to-admit-that-Israel-is-a-sick-society-that-needs-treatment-
     Jerusalem Post October 19th 2014.

PHOTO CREDITS: Marc Schulman and the I.D.F.